50 ideas

you really need to know

universe

Joanne Baker

Quercus

Contents

Introduction

Astronomy is one of the oldest and most profound of the sciences. Since our ancestors tracked the motions of the Sun and stars, what we have learned has radically altered our view of the place of humans in the universe. Each breakthrough has had social repercussions: Galileo were arrested in the 17th century for teaching the controversy that the Earth goes round the Sun. Demonstrations that our solar system is displaced from the heart of the Milky Way caused similar gasps of disbelief. And Edwin Hubble in the 1920s silenced the debaters when he discovered that the Milky Way is one of billions of galaxies scattered throughout a vast and swelling universe, 14 billion years old.

During the twentieth century technologies upped the pace of discovery. The century opened with gains in our understanding of stars and their fusion engines, paralleling our knowledge of nuclear power, radiation and the building of the atomic bomb. The years during and after the Second World War brought the development of radio astronomy, and the identification of pulsars, quasars and black holes. New windows on the universe were thrown open, from the cosmic microwave background radiation to the X-ray and gamma-ray sky, each frequency bringing its own discoveries.

This book takes a tour of astrophysics from a modern research perspective. The first chapters describe the great philosophical leaps in our understanding of the scale of the universe, whilst introducing the basics, from gravity to how a telescope works. The next set asks what we have learned about cosmology, the study of the universe as a whole – its constituent parts, history and evolution. Theoretical aspects of the universe, including relativity theory, black holes and multiverses, are then introduced. The last sections dissect in detail what we know about galaxies, stars and the solar system, from quasars and galaxy evolution to exoplanets and astrobiology. The pace of discovery is still high: perhaps in the next decades we will witness the next great paradigm shift – the detection of life beyond the Earth.

01 Planets

How many planets are there? A few years ago it was an easy question that anyone could answer – nine. Today, it is contentious. Astronomers have thrown a spanner in the works by discovering rocky bodies in the deep freeze of the outer solar system that rival Pluto, and by finding hundreds of planets around distant stars. Forced to rethink the definition of a planet, they now suggest there are eight bona fide planets in our solar system, plus a few dwarf planets like Pluto.

Since pre-history we've known that planets differ from stars. Planets, named after the Greek word for 'wanderer', migrate across the night sky through the unchanging backdrop of stars. From night to night the stars form the same patterns. Their constellations all slowly spin together about the north and south poles, each star etching a circle daily on the sky. But planets' positions shift a little relative to the stars each day, following a tilted path across the sky that is called the plane of the ecliptic. All the planets move within the same plane as they orbit the Sun, which is projected as a line on the sky.

The major planets other than Earth – Mercury, Venus, Mars, Jupiter and Saturn – have been known for millennia. They are easily visible to the naked eye, often outshining their stellar neighbours, and their contrary motions lent them mythical status. The arrival of telescopes in the 17th century generated more awe: Saturn was skirted by beautiful rings; Jupiter boasted a coterie of moons and Mars's surface was flecked by dark channels.

Planet X This heavenly certainty was shaken by the discovery of the planet Uranus in 1781 by the British astronomer William Herschel. Fainter and slower-moving than the other known planets, Uranus was originally thought to be a rogue star. It was Herschel's careful tracking that proved

timeline

350 BC	1543	1610	1781
Aristotle determines Earth is round	Copernicus publishes his heliocentric theory	Galileo Galilei discovers Jupiter's moons through telescope	William Herschel discovers Uranus

conclusively that it orbited the Sun, thus bestowing its planetary status. Herschel basked in fame because of his discovery, even courting King George III's favour by briefly naming it for the English monarch.

More discoveries were to come. Slight imperfections in Uranus's orbit led to predictions that it was being disturbed by another celestial body that lay beyond it. Several astronomers scoured the expected location, looking for a wandering interloper, and in 1846 Neptune was discovered by the Frenchman Urbain Jean Joseph Le Verrier, narrowly beating British astronomer John Couch Adams to establish the find.

Definition of planet

A 'planet' is a celestial body that: (a) is in orbit around the Sun, (b) has sufficient mass for its self-gravity to overcome rigid body forces so that it assumes a round shape, and (c) has cleared the neighbourhood around its orbit.

Then, in 1930, Pluto was confirmed. As was the case with Neptune, slight deviations in the expected movements of the outer planets suggested the presence of a further body – at the time called Planet X. Clyde Tombaugh at Lowell Observatory in the US spotted the object when comparing photographs of the sky taken at different times: the planet had given itself away by its motion. But it fell to a schoolgirl to name it. Venetia Burney from Oxford, in the UK, won a naming competition with her classics-inspired suggestion for Pluto, god of the underworld. Planet Pluto inspired a host of popular culture at the time, from the cartoon dog to the newly discovered element plutonium.

Pluto dethroned Our nine-planet solar system stood for another 75 years – until Michael Brown of Caltech and his collaborators discovered that Pluto was not alone. Having found a handful of sizeable objects not far from Pluto's orbit at the cold edge of the solar system, they happened upon one that was even larger than Pluto itself. They called it Eris. The astronomical community had a conundrum. Should Brown's discovery be recognised as a tenth planet?

'Like continents, planets are defined more by how we think of them than by someone's after-the-fact pronouncement.'

Michael Brown, 2006

1843–6	1930	1962	1992	2005
Neptune predicted and found by Adams and Le Verrier	Clyde Tombaugh discovers Pluto	First Mariner 2 images of Venus – surface of another planet	First extrasolar planet discovered	Brown discovers Eris

And what about the other icy bodies near Pluto and Eris? Pluto's status as a planet was called into question. The outer reaches of the solar system were littered with ice-smothered objects, of which Pluto and Eris were simply the largest. Moreover, rocky asteroids of similar size were known elsewhere, including Ceres, a 950-km-diameter asteroid that was found in 1801 between Mars and Jupiter during the search for Neptune.

In 2005 a committee of the International Astronomical Union, the professional organization of astronomers, met to decide Pluto's fate. Brown

WILLIAM HERSCHEL (1738–1822)

Born in Hanover, Germany, in1738, Frederick William Herschel emigrated to England in 1757 and earned a living as a musician. He developed a keen interest in astronomy, which he shared with his sister Caroline, whom he brought to England in 1772. The Herschels built a telescope to survey the night sky, cataloguing hundreds of double stars and thousands of nebulae. Herschel discovered Uranus and named it 'Georgium Sidum' in honour of King George III, who appointed him court astronomer. Herschel's other discoveries include the binary nature of many double stars, the seasonal variation of Mars's polar caps and the moons of Uranus and Saturn.

and some others wanted to protect the status of Pluto as culturally defined. In their view, Eris should also be considered a planet. Others felt that all the icy bodies beyond Neptune were not real planets. It came to a vote at a conference in 2006. What was decided was a new definition of a planet. Until then the concept was not pinned down. Some were bemused, saying that this was like asking for the precise definition of, say, a continent: if Australia is a continent, then what about Greenland? Where does Europe end and Asia begin? But the astrophysicists agreed a set of rules.

A planet is defined as a celestial body that orbits the Sun, is massive enough that its own gravity makes it round in shape, and has cleared the region around it. According to these rules, Pluto was not a planet, for it hadn't cleared other bodies from its orbit. Pluto and Eris were named dwarf planets, as was Ceres. Smaller bodies, apart from moons, remained unspecified.

Beyond the Sun This planetary definition was made for our own solar system. But it may be applied well beyond it. Today, several hundred planets are known to orbit stars other than the Sun. They were identified mainly by the subtle pulls they impart on their host stars. Most of these planets are massive gas giants, like Jupiter. But new spacecraft such as Kepler, launched in 2009, are vying to detect smaller planets around other stars that might be like Earth.

Another definition that has been called into question lately is that of a star. Stars are balls of gas, like the Sun, that are big enough to have ignited nuclear fusion in their cores. This energy makes the star shine. But it isn't obvious where the division is between planetary-sized balls of gas like Jupiter, and the smallest, dimmest stars, like brown dwarfs. Un-ignited stars and even free-floating planets may litter space.

> **'Maybe this world is another planet's hell.'**
> **Aldous Huxley**

the condensed idea
Planets stand out from the crowd

02 Heliocentrism

Although we now know the Earth and planets go round the Sun, this was not accepted until clues were amassed in the 17th century. It shattered our worldview: humans were not centrally placed in the universe, counter to the prevailing philosophy and religion at the time. Similar arguments about man's place in the cosmos rumble on, from creationist dogma to rational aspects of cosmology.

Early societies literally wanted the universe to revolve around them. In antiquity, models of the cosmos placed the Earth at the centre. Everything else radiated out from that. All the heavenly bodies, they imagined, were affixed to crystal spheres that spun about the Earth – causing the stars pinned upon them, or revealed through tiny holes within them, to circle the north and south celestial poles each night. The place of humans as key to the universe's machinations was assured.

Yet there were clues that this comforting model was incorrect; and they have puzzled natural philosophers for generations. The idea that the heavens revolve about the Sun rather than the Earth – a heliocentric model, after the Greek word *helios* for the Sun – was suggested by ancient Greek philosophers as early as 270 BC. Aristarchus of Samos was one who conveyed such ideas in his writings. After calculating the relative sizes of the Earth and the Sun, Aristarchus realized that the Sun was much the larger. It made more sense for the smaller Earth to move, rather than the bigger Sun.

Ptolemy in the second century used mathematics to predict the motions of the stars and planets. He did so reasonably well, but there were obvious patterns that his equations could not match. The most puzzling behaviour

timeline

270 BC

Ancient Greeks propose
heliocentric model

2nd century

Ptolemy adds epicycles to
explain retrograde motions

is that planets occasionally reverse direction – retrograde motion. Ptolemy, who imagined like those before him that the planets turned on vast circular wheels in the sky, invented an explanation by adding extra cogs to their orbits. He suggested that the planets rolled around smaller rings as they travelled along their main track, like a giant clockwork contraption. These superimposed 'epicycles' gave the appearance of the planets' occasional backward looping motions.

This idea of epicycles stuck, and was refined in later years. Philosophers were attracted to the idea that nature favoured perfect geometries. Nevertheless, as astronomers measured the planets' movements more accurately, their clockwork mathematical prescriptions increasingly failed to explain them. As their data improved, the discrepancies only grew.

Copernicus's model Heliocentric ideas were occasionally mooted over the centuries, but they were not taken seriously. The Earth-centred view prevailed by instinct, and alternative theories were thought arbitrary mind play. So it was not until the 16th century that the consequences of the Sun-centred model were fully developed. In his 1543 book *De Revolutionibus*, Polish astronomer Nicolaus Copernicus wrote down a mathematically detailed heliocentric model, explaining the planets' backwards motions as a projection of their movement around the Sun as viewed from a similarly circling Earth.

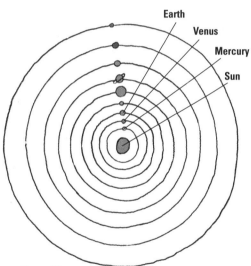

Earth
Venus
Mercury
Sun

❝Finally we shall place the Sun himself at the centre of the Universe.❞

Nicolaus Copernicus

1543
Copernicus publishes
heliocentric model

1609
Galileo discovers Jupiter's moons;
Kepler models orbits as ellipses

1633
Galileo put on trial for
teaching heliocentrism

NICOLAUS COPERNICUS (1473–1543)

Born in Torun, Poland, Copernicus trained to become a canon, taking classes in law, medicine, astronomy and astrology. He was fascinated by, but critical of, Ptolemy's ideas about the order of the universe, and instead worked out his own system where the Earth and planets rotate about the Sun.

Copernicus's work *De Revolutionibus Orbium Coelestium* (*On the Revolutions of the Heavenly Spheres*), published in March 1543 – just two months before he died – was groundbreaking in the establishment of the Sun-centred universe. Yet it is still far from the ideas of modern astronomy.

By challenging the universal pre-eminence of humans, Copernicus's model had consequences. The established Church and society favoured Ptolemy's Earth-centred view. Copernicus was cautious and delayed publication of his work until the year of his death. His posthumous argument was received and quietly put to one side. It fell to a more strident figure to carry the baton.

Galileo's conviction Italian astronomer Galileo Galilei notoriously challenged the Roman Catholic Church by championing heliocentrism. His audacity was backed up by observations he made through the then newly developed telescope. Peering into the heavens with greater clarity than his predecessors, Galileo found evidence that the Earth was not central to all. Jupiter had moons orbiting it and Venus had phases like the Moon. He published these discoveries in his 1610 book *Sidereus Nuncius*, or *Starry Messenger*.

Confident in his Sun-centred view, Galileo defended his argument in a letter to the Grand Duchess Christina. Having claimed that it was the Earth's rotation that gave the appearance of the Sun moving across the sky, he found himself summoned to Rome. The Vatican conceded that his observations were true, because Jesuit astronomers saw the same things through their telescopes. But the Church refused to accept Galileo's theory, stating that it was just a hypothesis and could not be taken literally, however appealing it was in its simplicity. In 1616 the Church banned Galileo from teaching heliocentrism, and precluded that he 'hold or defend' that contentious idea.

> **It is surely harmful to souls to make it a heresy to believe what is proved.**
> **Galileo Galilei**

Kepler's reason Meanwhile, a German astronomer was also working on the mathematics of planetary motions. Johannes Kepler published his analysis of the path of Mars in the book *Astronomia nova* (1609), in the same year that Galileo took up his telescope. Kepler found that an ellipse rather than a circle gave a better description of the red planet's orbit about the Sun. In freeing himself from perfect circles, he moved beyond Copernicus's model and improved the predictions for planetary motions. Although now considered a basic law of physics, Kepler's vision was advanced for its time and took a long time to be accepted. Galileo, for one, took no notice.

Although restricted, Galileo remained certain that his Sun-centred explanation was true. Asked by Pope Urban VIII to write a balanced account of both sides, in *Dialogue of the two world systems* Galileo upset the pontiff by expressing a bias for his own view over that of the Church. The Vatican again summoned him to Rome and put him on trial in 1633 for breaking his ban. Galileo was placed under house arrest for the rest of his life, dying in 1642. A formal apology from the Vatican was only made four centuries later, in the run-up to the publication anniversary of his contentious book.

Gradual acceptance Evidence that the heliocentric view of the solar system was correct accumulated steadily over the centuries. Kepler's mechanics of orbits was found to hold and also influenced Newton's theory of gravity. As further planets were discovered, the fact that they orbited the Sun was obvious. Man's place at the centre of things is no longer tenable.

the condensed idea
The Sun's in the centre

03 Kepler's laws

Johannes Kepler looked for patterns in everything. Peering at astronomical tables describing the looped motions of Mars projected on the sky, he discovered three laws that govern the orbits of the planets. He described how planets follow elliptical orbits and how more distant planets orbit more slowly around the Sun. As well as transforming astronomy, Kepler's laws laid the foundations for Newton's law of gravity.

As the planets orbit around the Sun, the closest ones move more quickly around it than those further away. Mercury circles the Sun in just 80 Earth days. If Jupiter travelled at the same speed it would take about 3.5 Earth years to complete an orbit when, in fact, it takes 12. As all the planets sweep past each other, when viewed from the Earth some appear to backtrack as the Earth moves forwards past them. In Kepler's time these 'retrograde' motions were a major puzzle. It was solving this puzzle that gave Kepler the insight to develop his three laws of planetary motion.

Patterns of polygons Kepler was a German mathematician who lived in the late 16th and early 17th centuries, a time when astrology was taken very seriously and astronomy as a physical science was still in its infancy. Religious and spiritual ideas were just as important as observation in revealing nature's laws. A mystic who believed that the underlying structure of the universe was built from perfect geometric forms, Kepler devoted his life to trying to tease out the patterns of imagined perfect polygons hidden in nature's works.

Kepler's work came almost a century after Polish astronomer Nicolaus Copernicus proposed that the Sun lies at the centre of the universe and the Earth orbits the Sun, rather than the other way around. At first, Kepler

timeline

*c.*580 BC	*c.*150	1543
Pythagoras states that planets orbit in perfect crystalline spheres	Ptolemy suggests epicycles for retrograde motion	Copernicus proposes that planets orbit the Sun

adopted Copernicus's heliocentric idea, believing that the planets moved round the Sun in circular orbits. He envisaged a system in which the planets' orbits lay within a series of nested crystal spheres spaced according to mathematical ratios. These scalings were derived from the sizes of a series of polygons with increasing numbers of sides that fitted within the spheres. The idea that nature's laws followed basic geometric ratios had originated with the ancient Greeks.

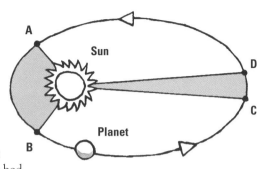

Trying to model the orbits of the planets to support his geometric ideas, Kepler used the most accurate data available – intricate tables of the planets' motions on the sky, painstakingly prepared by Tycho Brahe. In these columns of numbers Kepler saw patterns that made him revise his thoughts and suggested his three laws.

Kepler got his breakthrough by disentangling the retrograde motions of Mars. Every so often the red planet would reverse its path on the sky and perform a small loop. Copernicus had modelled the loops by adding to the main orbit small extra turns from circular 'epicycles' superimposed on it. But Kepler found that his accurate new measurements failed to match those

> **It suddenly struck me that that tiny pea, pretty and blue, was the Earth. I put up my thumb and shut one eye, and my thumb blotted out the planet Earth. I didn't feel like a giant. I felt very, very small.**
>
> **Neil Armstrong**

1576	**1609**	**1687**	**2009**
Tycho Brahe maps the planets' positions	Kepler publishes theory of orbits as ellipses	Newton explains Kepler's laws with gravity	NASA launches Kepler satellite to detect planets around other stars

Kepler's Laws

First law: planetary orbits are elliptical with the Sun at one focus. Second law: a planet sweeps out equal areas in equal times as it orbits the Sun. Third law: the orbital periods scale with ellipse size, such that the period squared is proportional to the major axis length cubed.

predictions. Seeking another explanation he had the brainwave that the backward loops would fit if the planets' orbits were elliptical around the Sun and not circular as had been thought. Ironically this meant that nature did not follow perfect shapes, as Kepler had first imagined, but he was brave enough to accept the evidence and change his mind.

Orbits In Kepler's first law, he noted that the planets move in elliptical orbits with the Sun at one of the two foci of the ellipse. His second law describes how quickly a planet moves around its orbit. As the planet progresses along its path, it sweeps out an equal area segment in an equal time. The segment is measured using the angle drawn between the Sun and the planet's two positions (AB or CD), like a slice of pie. Because the orbits are elliptical, when the planet is close to the Sun it needs to cover a larger distance to sweep out the same area than when it is further away. So the planet moves faster near the Sun than when it is distant. Kepler's second law ties its speed with its distance from the Sun. Although he didn't realize it at the time, this behaviour is ultimately due to gravity accelerating the planet faster when it is near the Sun's mass.

Kepler's third law goes one step further again and tells us how the orbital periods scale up for different-sized ellipses at a range of distances from the Sun. It states that the squares of the orbital periods are proportional to the cube power of the longest axis of the elliptical orbit. The larger the elliptical orbit, the slower the period of time taken to complete an orbit. So planets further from the Sun orbit more slowly than nearby planets. Mars takes nearly two Earth years to go around the Sun, Saturn 29 years and Neptune 165 years.

We are just an advanced breed of monkeys on a minor planet of a very average star. But we can understand the universe. That makes us something very special.

Stephen Hawking

> **‟I measured the skies, now the shadows I measure, Sky-bound was the mind, earth-bound the body rests.”**
>
> **Kepler's epitaph**

In these three laws Kepler managed to describe the orbits of all the planets in our solar system. His laws apply equally to any body in orbit around another, from comets, asteroids and moons in our solar system to planets around other stars and even artificial satellites whizzing around the Earth. Four centuries after he proposed them, his laws are still a mainstay of physics. Moreover Kepler was ahead of his time in that he was one of the first to use the scientific methods that we use today – to make and analyse observations to test theories.

Kepler succeeded in unifying the principles into geometric laws but he did not know why these laws held. He believed that they arose from the underlying geometric patterns of nature. It took Newton to unify these laws into a universal theory of gravity.

JOHANNES KEPLER 1571–1630

Johannes Kepler liked astronomy from an early age, recording in his diary a comet and a lunar eclipse before he was ten. While teaching at Graz, he developed a theory of cosmology that was published in the *Mysterium Cosmographicum* (*The Sacred Mystery of the Cosmos*). He later assisted astronomer Tycho Brahe at his observatory outside Prague, inheriting his position as Imperial Mathematician in 1601. There Kepler prepared horoscopes for the emperor and analysed Tycho's astronomical tables, publishing his theories of noncircular orbits, and the first and second laws of planetary motion, in *Astronomia Nova* (*New Astronomy*). In 1620, Kepler's mother, a herbal healer, was imprisoned as a witch and only released through Kepler's legal efforts. However, he managed to continue his work and the third law of planetary motion was published in *Harmonices Mundi* (*Harmony of the Worlds*).

the condensed idea
Law of the worlds

04 Newton's law of gravitation

Isaac Newton made a giant leap when he connected the motions of cannonballs to the movements of the planets, thus linking heaven and Earth. His law of gravitation remains one of the most powerful ideas of physics, explaining motions both in our world and across the universe. Newton argued that all bodies attract each other through the force of gravity, and that the strength of that force drops off with distance squared.

The idea of gravity supposedly came to Newton when he saw an apple fall from a tree. We don't know if this story is true, but Newton stretched his imagination from earthly to heavenly motions to work out his law of gravitation. He perceived that objects were attracted to the ground by some accelerating force. If apples fall from trees, what if the tree were even higher? What if it reached to the Moon? Why doesn't the Moon fall to the Earth like an apple? he wondered.

All fall down Newton's answer lay first in his laws of motion linking forces, mass and acceleration. A ball blasted from a cannon travels a certain distance before falling to the ground. What if it were fired more quickly? Then it would travel further. If it was fired so fast that it travelled far enough in a straight line that the Earth curved away beneath it, where would it fall? Newton realized that it would be pulled towards Earth but would then follow a circular orbit, just like a satellite constantly being pulled but never reaching the ground.

timeline

350 BC	1609
Aristotle discusses why objects fall	Kepler reveals the laws of planetary orbits

ISAAC NEWTON 1643–1727

Isaac Newton was the first scientist to be honoured with a knighthood in Britain. Despite being 'idle' and 'inattentive' at school, and an unremarkable student at Cambridge University, he flourished suddenly when plague closed the university in the summer of 1665. Returning to his home in Lincolnshire, he devoted himself to mathematics, physics and astronomy, and even laid the foundations for calculus. There he produced early versions of his three laws of motion and deduced the inverse square law of gravity. After this remarkable outburst of ideas, Newton was elected to the Lucasian Chair of Mathematics in 1669 at just 27 years old. Turning his attention to optics, he discovered with a prism that white light was made up of rainbow colours, quarrelling famously with Robert Hooke and Christiaan Huygens over the matter. Newton wrote two major works, *Philosophiae naturalis principia mathematica*, or *Principia*, and *Opticks*. Late in his career, he became politically active. He defended academic freedom when King James II tried to interfere in university appointments, and entered Parliament in 1689. A contrary character, on the one hand desiring attention and on the other being withdrawn and trying to avoid criticism, Newton used his powerful position to fight bitterly against his scientific enemies and remained a contentious figure until his death.

When Olympic hammer-throwers spin on their heels, it is the pull on the string that keeps the hammer rotating. Without this pull the hammer would fly off in a straight line, just as it does on its release. It's the same with Newton's cannonball – without the centrally directed force tying the projectile to Earth, it would fly off into space. Thinking further, Newton reasoned that the Moon also hangs in the sky because it is held by the invisible tie of gravity. Without gravity, it too would fly off into space.

Inverse square law Newton then tried to quantify his predictions. After exchanging letters with his contemporary Robert Hooke, he showed that gravity follows an inverse square law – the strength of gravity decreases by the square of the distance from a body. So if you travel twice some distance

'Gravity is a habit that is hard to shake off.'

Terry Pratchett

1687	**1905**	**1915**
Newton's *Principia* is published	Einstein publishes the special theory of relativity	Einstein publishes the general theory of relativity

from a body, its gravity is four times less; the gravity exerted by the Sun would be four times less for a planet in an orbit twice as far from it as the Earth, or a planet three times distant would experience gravity nine times less.

Newton's inverse square law of gravity explained in one equation the orbits of all the planets as described in the three laws of Johannes Kepler (see p.14). Newton's law predicted that the planets travelled more quickly near the Sun as they followed their elliptical paths. A planet feels a stronger gravitational force from the Sun when it travels close to it, which makes it speed up. As its speed increases, the planet is thrown away from the Sun again, gradually slowing back down. Thus Newton pulled together all the earlier work into one profound theory.

Universal law Generalizing boldly, Newton then proposed that his theory of gravity applied to everything in the universe. Any body exerts a gravitational force in proportion to its mass, and that force falls off with distance squared. So any two bodies attract each other, but because gravity is a weak force we only really observe this for very massive bodies, such as the Sun, Earth and planets.

If we look closer, though, it is possible to see tiny variations in the local strength of gravity on the surface of the Earth. Because massive mountains and rocks of differing density can raise or reduce the strength of gravity near them, it is possible to use a gravity meter to map out geographic terrains and to learn about the structure of the Earth's crust. Archaeologists also sometimes use tiny gravity changes to spot buried settlements. Recently, scientists have used gravity-measuring space satellites to record the (decreasing) amount of ice covering the Earth's poles and also to detect changes in the Earth's crust following large earthquakes.

Acceleration

On the surface of the Earth the acceleration of a falling body due to gravity, g, is 9.81 metres per second per second.

❝**Every object in the universe attracts every other object along a line of the centres of the objects, proportional to each object's mass, and inversely proportional to the square of the distance between the objects.**❞

Isaac Newton

Back in the 17th century, Newton poured all his ideas on gravitation into one book, *Philosophiae naturalis principia mathematica*, known as the *Principia*. Published in 1687, it is still revered as a scientific milestone. Newton's universal gravity explained the motions not only of planets and moons but also of projectiles, pendulums and apples. He explained the orbits of comets, the formation of tides and the wobbling of the Earth's axis. This work cemented his reputation as one of the great scientists of all time.

Relativity Newton's universal law of gravitation has endured for hundreds of years and still today gives a basic description of the motion of bodies. However, science does not stand still, and twentieth-century scientists have built upon its foundations, notably Einstein with his theory of general relativity. Newtonian gravity still works well for most objects we see and for the behaviour of planets, comets and asteroids in the solar system that are spread over large distances from the Sun where gravity is relatively weak. Although Newton's law of gravitation was powerful enough to predict the position of the planet Neptune, discovered in 1846 at the expected location beyond Uranus, it was the orbit of another planet, Mercury, that required physics more advanced than that of Newton. Thus general relativity is needed to explain situations where gravity is very strong, such as close to the Sun, stars and black holes.

the condensed idea
Mass attraction

05 Newton's theory of optics

Astronomers reveal many of the universe's secrets by exploiting the physics of light. Isaac Newton was one of the first to try to understand its nature. Passing white light through a glass prism, he found that it split into rainbow hues and showed that the colours were embedded in the white light rather than imprinted by the prism. Today we know that visible light is one segment of a spectrum of electromagnetic waves, stretching from radio waves to gamma rays.

Shine a beam of white light through a prism and the emerging ray spreads out into a rainbow of colours. Rainbows in the sky appear in the same way: sunlight is split by water droplets into the familiar spectrum of hues: red, orange, yellow, green, blue, indigo and violet.

Experimenting with light beams and prisms in his rooms in the 1660s, Isaac Newton demonstrated that light's many colours could be mixed together to form white light. Colours were the base units rather than being made by later mixing or by the prism glass itself, as had been thought. Newton separated beams of red and blue light and showed that these single colours were not split further if they were passed through more prisms.

Light waves Experimenting further, he concluded that light behaves in many ways like water waves. Light bends around obstacles in a similar way to sea waves around a harbour wall. Light beams can also be added together to reinforce or cancel out their brightness, as overlapping water waves do.

> **'Light brings us news of the Universe.'**
>
> **Sir William Bragg**

timeline

1672	1678
Newton explains rainbow	Christiaan Huygens publishes a wave theory of light

In the same way that water waves are large-scale motions of invisible water molecules, Newton believed that light waves were ultimately ripples of minuscule light particles, or 'corpuscles', which were even smaller than atoms.

What Newton did not know, something that was not discovered until centuries later, is that light waves are electromagnetic waves – waves of coupled electric and magnetic fields – and not the reverberation of solid particles. When the electromagnetic wave behaviour of light was discovered, Newton's corpuscle idea was put on ice. It was resurrected, however, in a new form when Albert Einstein showed that light may also behave sometimes like a stream of particles that can carry energy but have no mass.

Across the spectrum The different colours of light reflect the different wavelengths of these electromagnetic waves. Wavelength is the measured distance between consecutive crests of a wave. As it passes through a prism, the white light separates into many hues

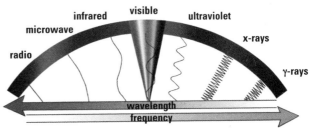

because each hue is deflected to a different degree by the glass. The prism bends the light waves by an angle that depends on the wavelength of light, where red light is bent least and blue most, to produce the rainbow colour sequence. The spectrum of visible light appears in order of wavelength, from red with the longest through green to blue with the shortest.

What lies at either end of the rainbow? Visible light is just one part of the electromagnetic spectrum. It is important to us because our eyes have developed to use this sensitive part of the sequence. As the wavelengths of visible light are on roughly the same scale as atoms and molecules (hundreds of billionths of a metre), the interactions between light and atoms in a material are large. Our eyes have evolved to use visible light

1839	**1873**	**1895**	**1905**
Alexandre Becquerel observes the photoelectric effect	James Clerk Maxwell's equations show light is an electromagnetic wave	Wilhelm Roentgen discovers X-rays	Einstein shows light can behave as particles in some circumstances

because it is very sensitive to atomic structure. Newton was fascinated by how the eye worked; he even stuck a darning needle round the back of his own eye to see how pressure affected his perception of colour.

Beyond red light comes infrared, with wavelengths of millionths of a metre. Infrared rays carry the Sun's warmth and are also collected by night-vision goggles to 'see' the heat from bodies. Longer still are microwaves, with millimetre-to-centimetre wavelengths, and radio waves, with wavelengths of metres and longer. Microwave ovens use microwave electromagnetic rays to spin the water molecules within food, heating them up. At the other end of the spectrum, beyond blue, comes ultraviolet light. This is emitted by the Sun and can damage our skin, although much of it is stopped by the Earth's ozone layer. At even shorter wavelengths are X-rays – used in hospitals because they travel through human tissue – and at the smallest wavelengths are gamma rays. Astronomers look at the universe at all these wavelengths.

Photons But light doesn't always behave like a wave – Newton was partly right. Light rays do carry energy that is delivered in tiny packets, called photons, which have no mass and travel at the speed of light. This was realized by Albert Einstein, who saw that blue and ultraviolet light shone on to a metal set up an electric current – the photoelectric effect. The currents are generated when metals are illuminated by blue or ultraviolet light, but not red light. Even a bright beam of red light fails to trigger a current. Charge flows only when the light's frequency exceeds

Matter waves

In 1924, Louis-Victor de Broglie suggested the converse idea that particles of matter could also behave as waves. He proposed that all bodies have an associated wavelength, implying that particle–wave duality was universal. Three years later the matter-wave idea was confirmed when electrons were seen to diffract and interfere just like light. Physicists have now also seen larger particles behaving like waves, such as neutrons, protons and recently even molecules including microscopic carbon footballs or 'bucky balls'. Bigger objects, like ball bearings and badgers, have minuscule wavelengths, too small to see, so we cannot spot them behaving like waves. A tennis ball flying across a court has a wavelength of $10–34$ metres, much smaller than a proton's width (10^{-15} m).

'Nature and nature's laws lay hid in night; God said "Let Newton be" and all was light.'

Alexander Pope (Newton's epitaph)

some threshold, which varies for different metals. The threshold indicates that a certain amount of energy needs to be built up before the charges can be dislodged.

In 1905, Einstein came up with a radical explanation. It was this work, rather than relativity, that won him the Nobel Prize in 1921. Rather than bathing the metal with continuous light waves, he suggested that individual photon bullets hit electrons in the metal into motion to produce the photoelectric effect. Because each photon carries a certain energy, scaling with its own frequency, the bumped electron's energy also scales with the light's frequency.

A photon of red light (with a low frequency) cannot carry enough energy to dislodge an electron, but a blue photon (light with a higher frequency) has more energy and can set it rolling. An ultraviolet photon has more energy still, so it can slam into an electron and donate even more speed. Turning up the brightness of light changes nothing; it doesn't matter that you have more red photons if each is incapable of shifting the electrons. It's like firing ping-pong balls at a weighty sports utility vehicle. Einstein's idea of light quanta was unpopular at first, but the climate altered when experiments showed his wacky theory to be true. They confirmed that the energies of the liberated electrons scaled proportionally with the frequency of light.

Wave-particle duality Einstein's proposal raised the uncomfortable idea that light was both a wave and a particle, called wave-particle duality. Physicists are still struggling with this tension. Today, we even understand that light seems to know whether to behave as one or the other under different circumstances. If you set up an experiment to measure its wave properties, such as passing it through a prism, it behaves as a wave. If instead you try to measure its particle properties it is similarly obliging. It is truly both.

the condensed idea
Beyond the rainbow

06 The telescope

Modern astronomy began with the invention of the telescope in the 17th century. It opened up the solar system to view, revealing Saturn's rings and allowing the discovery of the outer planets. Telescope observations were crucial in confirming that the Earth orbits the Sun. Eventually it gave access to the entire visible universe.

Galileo was famously one of the first astronomers to look through a telescope, using its magnification in 1609 to discover four of Jupiter's moons, Venus's phases and craters of the Moon. Yet he was just following the fashion.

No individual is credited with the invention of the telescope. Dutchman Hans Lipperhey was one of the first to attempt to patent a telescope design in 1608, but he was unsuccessful because the concept was already widely known. The magnifying power of transparent material with curved surfaces was well recognized; and the lentil-shaped 'lens' had been in use in magnifying glasses and spectacles at least since the 13th century. Records show that telescopes had been built and used to look at the Moon in the mid 16th century, but developments in glass-making meant that quality instruments only became widespread in the 17th century. Good lenses produced clean images, even of faint heavenly bodies.

Magnifying power How does a telescope work? The simplest version uses two lenses slotted at either end of a tube. The first lens squeezes rays of light inwards so that the eye perceives them as coming from a larger source. The second lens acts as an eyepiece, making the light rays parallel again before they enter the eye so that they can be focused.

timeline

1609
Galileo used a telescope for astronomy

1668
Newton builds a reflecting telescope

The bending of rays by the lens is called refraction. Light travels more slowly in denser materials, such as glass, compared with air. This explains the mirage of a puddle on a hot road. Rays from the sky bend to skim the road's surface because light changes speed in the layer of hot air lying just above the sun-baked asphalt. Hot air is less dense than cooler air, so the light bends away from the vertical and we see the sky's reflection on the tarmac, looking like a wet puddle.

> **We see past time in a telescope and present time in a microscope. Hence the apparent enormities of the present.**
>
> **Victor Hugo**

The angle by which a ray bends is related to the relative speeds at which it travels in the two materials – technically, the ratio of the speeds gives the ratio of the sine of the incident angles, measured from the vertical. So for a ray leaving air for glass, and other dense substances, it is bent inwards and its path becomes steeper.

Refractive index Light travels at a whopping 300 million metres per second in empty space. The ratio of its velocity in a denser material such as glass to that in a vacuum is called the refractive index of the material. A vacuum has, by definition, a refractive index of 1; something with a refractive index of 2 would slow light to half its speed in free space. A high refractive index means that light bends a lot as it passes through the substance.

Refractive index is a property of the material itself. Materials can be designed to possess specific refractive indices, which may be useful, for example, when designing telescopes or lenses for glasses to correct problems with someone's vision. The power of lenses and prisms depends on their refractive index; high-power lenses have high refractive indices.

Refracting telescopes with two lenses have drawbacks. The final image appears upside down, because the light rays cross over before they reach the eyepiece.

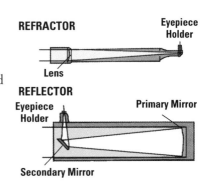

REFRACTOR — Eyepiece Holder — Lens

REFLECTOR — Eyepiece Holder — Primary Mirror — Secondary Mirror

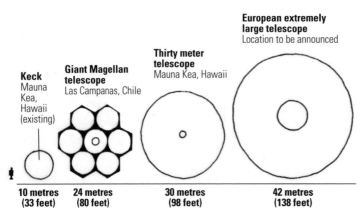

Keck
Mauna
Kea,
Hawaii
(existing)

Giant Magellan telescope
Las Campanas, Chile

Thirty meter telescope
Mauna Kea, Hawaii

European extremely large telescope
Location to be announced

10 metres (33 feet)   **24 metres (80 feet)**   **30 metres (98 feet)**   **42 metres (138 feet)**

For astronomy this usually isn't a problem, as a star looks much the same upside down. It can be corrected by including a third lens to invert the image again, but then the telescope can become long and cumbersome. Second, and most problematically, refracting telescopes produce blurred colour images. Because different wavelengths of light are refracted by different amounts – blue light waves are bent more than red light waves – the colours separate out and the final image loses clarity. New types of lenses that are available today can minimize this, but their size is limited.

Reflecting telescope To solve these problems, Newton invented the reflecting telescope. Using a curved mirror rather than a lens to bend the light, he essentially folded the telescope in half, making it easier to handle. His design also avoided the differential blurring because the mirrored surface reflects all colours of light in the same way. However, mirror silvering techniques were not advanced in Newton's day, and it took centuries for the design to be perfected.

Today, most professional astronomical telescopes use a giant mirror, rather than a lens, to collect celestial light and bounce it back to the eyepiece. The size of the mirror dictates how much light can be collected – a big area lets you see very faint objects. The mirrors in modern optical telescopes can be the size of a room – the largest currently in use, such as those in the twin giant Keck telescopes on Mauna Kea, Hawaii, are around 10 metres across. Even bigger ones up to 100 metres in diameter are planned in the coming decades.

Very large mirrors are difficult to build. They become so heavy that their shape distorts when the telescope tilts to scan the sky. Clever construction methods are needed to make them as light as possible. Some are built in many segments; others are carefully spun so that they are thin yet accurately sculpted. An alternative solution, called 'adaptive optics', is to constantly correct the mirror's shape using a network of tiny pistons glued underneath to push up the surface when it sags.

Twinkling stars Beyond the telescopes themselves, the clarity of astronomical images is degraded by turbulence in our atmosphere. On even the clearest night, stars twinkle. Those near the horizon twinkle more than those overhead. They do so because of pockets of air moving in front of them. Astronomers call the blurring of stars by our atmosphere 'seeing'. The size of the optical components in the telescope also gives an absolute limit to the concentration of starlight due to another behaviour of light, diffraction – the bending of light rays around the edge of a lens, aperture or mirror.

> **'Where there is an observatory and a telescope, we expect that any eyes will see new worlds at once.'**
>
> **Henry David Thoreau**

To get the sharpest images of stars and planets, astronomers choose special locations for their telescopes. On the surface of the Earth, they build them on high sites where the air is thin, such as mountains, and airflow is smooth, such as near the coast. The best sites include the Chilean Andes and Hawaii's volcanic peaks. The ultimate site is in space, where there is no atmosphere. The deepest images ever taken of the universe have been made by the orbiting Hubble Space Telescope.

Telescopes can operate at wavelengths other than the visible range. Infrared light, or heat, can be detected with instruments that are like night-vision goggles mounted on telescopes, as long as the equipment is kept cool. Because of their very short wavelengths, X-rays are best pursued in space using satellites with reflective optics. Even radio waves can be detected with large single dishes, such as the one at Arecibo, which has appeared in James Bond films, or arrays of many smaller antennae, such as the Very Large Array in New Mexico, which featured in *Contact*. Perhaps the ultimate telescope is the Earth itself – fundamental particles whiz through it every day, and physicists have placed traps to try to catch them as they do.

the condensed idea
Light-bending magnification

07 Fraunhofer lines

Within the spectrum of starlight lies a chemical fingerprint. Dark or bright lines signpost specific wavelengths that are absorbed or emitted by scorching gas in a star's atmosphere. First noticed in light from the Sun, these atomic markers are a powerful tool for astronomical detective work. They have revealed the chemical make-up of stars and galaxies as well as the motions of celestial bodies, and the expansion of the universe.

If you pass sunlight through a prism, the rainbow spectrum that emerges is striped by a series of dark lines, like a bar code. They mark particular wavelengths of light that are chopped out because they are absorbed by gases in the Sun's atmosphere. Each line corresponds to a particular chemical element seen in various states and energies, from neutral atoms to excited ions. By mapping the pattern of the lines you can work out the chemistry of the Sun.

Although spotted by English astronomer William Hyde Wollaston in 1802, absorption lines in the solar spectrum were first examined in detail in 1814 by top German lens-maker Joseph von Fraunhofer, after whom they are named. He was able to discern more than 500 lines; with modern equipment we can detect many thousands of them.

Unique chemistry German chemists Gustav Kirchhoff and Robert Bunsen worked out in their laboratory in the 1850s that each element gives rise to a unique ladder pattern of absorption lines. In the Sun, hydrogen is

timeline

1802	1814
Wollaston sees dark lines in the Sun's spectrum	Fraunhofer measures hundreds of lines

the most common element, and the solar spectrum also shows absorption from many others, including helium, carbon, oxygen, sodium, calcium and iron. Each has its own absorption line bar code.

The light from other stars also carries chemical imprints. The study of spectral chemistry, known as spectroscopy, is an especially powerful technique in astronomy because it reveals the material that makes up stars and also nebulae, planetary atmospheres and distant galaxies. Astronomers can't bring stars and galaxies into their laboratory, or travel to them, so they must resort to remote observations and clever techniques.

Sometimes these lines appear bright rather than dark: these are known as emission lines. Very bright sources, such as the hottest stars and luminous quasars, are so energetic that their gases try to cool down by releasing photons at these characteristic wavelengths rather than absorbing them. Fluorescent lights also emit a series of bright lines that correspond to the wavelengths of excited atoms in the gas in the tube, such as neon.

'It is the stars, The stars above us, govern our conditions.'

William Shakespeare

Gratings To split light into its constituent wavelengths, devices known as gratings are often used. Rather than glass prisms, which are bulky and limited in the amount by which they can bend light by their refractive index, a

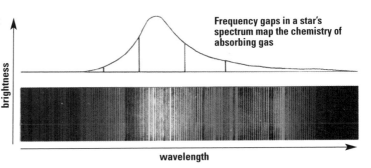

Frequency gaps in a star's spectrum map the chemistry of absorbing gas

brightness

wavelength

component with a row of parallel narrow slits cut into it is inserted into the light beam. Fraunhofer made the first grating from aligned wires.

1842	1859	1912
Doppler explains shifted spectral lines	Kirchhoff and Bunsen discover spectroscopy in the lab	Vesto Slipher discovers that galaxies are redshifted

Gratings work because of the wave properties of light. Light passing through each slit of the grid is spread out by diffraction, the amount being proportional to the wavelength of the light but inversely proportional to the width of the slit. Very narrow slits spread the light out more broadly; and red light is diffracted more than blue light.

The multiple slits combine the light further using another property, interference –where peaks and troughs of light waves either reinforce or cancel each other out to create a superposed pattern of light and dark fringes. Within each of these fringes the light is split even more finely; again scaling with wavelength but this time inversely proportional to the distance between the slits. By controlling the number of divisions, their separation and width, astronomers can control the amount by which the light is spread out, and the finesse with which they can probe absorption and emission lines. Gratings are therefore much more powerful and versatile than prisms.

A simple grating can be made from a photographic slide with slits etched into it; they are sometimes sold in science museum shops. If you put one up to a neon light, you will see the bar-code wavelengths of the hot gas spread out in front of your eye.

Diagnostics Spectral lines are more than chemical indicators. Because each line corresponds to a particular atomic state, their wavelengths are very well known from laboratory experiments. Each line's characteristic

JOSEPH VON FRAUNHOFER 1787–1826

Born in Bavaria in 1787, Joseph von Fraunhofer rose from humble beginnings to be a world-class optical glass-maker. After he was orphaned at the age of 11, he became an apprentice glass-maker. When the workshop he was apprenticed to collapsed in 1801, he was rescued by a Bavarian prince who saw to it that he was allowed to study. Learning his speciality at a top monastery, Fraunhofer became a world-renowned maker of optical glass and instruments. His scientific career was illustrious; he became director of the Optical Institute, a nobleman and an honorary citizen of Munich. But like many glass-makers of his day, he died young, at the age of 39, due to poisoning by heavy metal vapours.

> **'Why I came here, I know not; where I shall go it is useless to inquire – in the midst of myriads of the living and the dead worlds, stars, systems, infinity, why should I be anxious about an atom?'**
> **Lord Byron**

energy originates in the structure of the atom. Although in reality they are much more complicated and ephemeral, atoms can be thought of simply in a similar way to our solar system. The nucleus, comprised of weighty protons and neutrons, is like the Sun; and the electrons are like the planets. Absorption and emission lines arise when planets move from one orbit to another, when energy in the form of a photon is either applied or taken out.

Absorption happens when a photon of the right energy comes in and knocks an electron to a higher orbit; emission when an electron falls to a lower orbit by donating the extra energy to the photon. The energies required to move between orbits are precisely defined, and depend on the type and state of the atom. In very hot gases, the outer electrons may be stripped off altogether – the atoms are said to be ionized.

Because of their origin in fundamental physics, spectral lines are sensitive to many aspects of the physics of the gas. Its temperature can be deduced from the broadening of the lines, such that hotter gas produces broader lines. Ratios of spectral line strengths give further information, such as the degree of ionization of the gas.

Another use for spectral lines is to measure the motions of celestial bodies. The wavelength of a particular line is accurately known, so any slight shifts in that line may indicate movement of the source. If the entire star is moving away from us, then its spectrum shifts to the red due to the Doppler effect (see p.32–35); towards us then it shifts to the blue. The amount of the shift can be measured by looking at spectral lines. On a grander scale, these 'redshifts' have even given away the expansion of the universe.

the condensed idea
Bar codes in the stars

08 Doppler effect

We've all heard the drop in pitch of an ambulance siren's wail as it speeds past. Waves coming from a source that is moving towards you arrive squashed together and so seem to have a higher frequency. Similarly, waves become spread out and so take longer to reach you from a source that is receding, resulting in a frequency drop. This is the Doppler effect. It has been used to measure speeding cars, blood flow and – as redshift – the motions of stars and galaxies in the universe.

The Doppler effect was proposed by Austrian mathematician and astronomer Christian Doppler in 1842. It arises because of the motion of the emitting vehicle relative to you, the observer. As the vehicle approaches, its sound waves pile up, the distance between each wavefront is squashed and the sound gets higher. As it speeds away, the wavefronts consistently take a little longer to reach you, the intervals get longer and the pitch drops. Sound waves are pulses of compressed air.

To and fro Imagine that someone on a moving train is throwing balls to you continually at a frequency of one ball every three seconds, prompted by their wristwatch timer. If they are motoring towards you, it will always take a little less than three seconds for the balls to arrive because they are launched a little closer to you each time, so the rate will seem quicker to the catcher. Similarly, as the train moves away, the balls take slightly longer to arrive, travelling a little extra distance each throw, so their arrival frequency is lower. If you could measure that shift in timing with your own watch, then you could work out the speed of the thrower's train. The Doppler effect applies to any objects moving relative to one another. It would be the same if you were moving on the train and the ball thrower was

timeline

1842

Doppler presents his paper
on colour shift in starlight

Extra-solar planets

More than 200 planets have been discovered orbiting around stars other than our Sun. Most are gas giants similar to Jupiter though orbiting much closer to their central stars, but a few possible rocky planets, similar to the Earth in size, have been spotted. About one in ten stars have planets, and this has fuelled speculation that some may even harbour forms of life. The great majority of planets have been found by observing the gravitational tug of the planet on its host star. Planets are tiny compared to the stars they orbit, so it is hard to see them against their star's glare.

But the mass of a planet swings the star around a little, and this wobble can be seen as a Doppler shift in the frequency of a characteristic feature in the spectrum of the star. The first extra-solar planets were detected around a pulsar in 1992 and around a normal star in 1995. Their detection is now routine, but astronomers are still seeking Earth-like solar systems and trying to figure out how different planetary configurations occur. New space observatories, such as Nasa's Kepler spacecraft, which was launched in 2009, are expected to identify many Earth-like planets.

standing still on a stationary platform. As a way of measuring speed, the Doppler effect has many applications. It is used in medicine to measure blood flow and also in roadside radars that catch speeding drivers.

Motion in space Doppler effects also appear frequently in astronomy, showing up wherever there is moving matter. For example, light coming from a planet orbiting a distant star would show Doppler shifts. As the planet moves towards us, the frequency rises, and as it spins away, its light

❝Perhaps when distant people on other planets pick up some wavelength of ours all they hear is a continuous scream.❞
Iris Murdoch

1912
Vesto Slipher measures redshifts of galaxies

1992
First detection of an extra-solar planet by the Doppler method

CHRISTIAN DOPPLER 1803–53

Christian Doppler was born into a family of stonemasons in Salzburg, Austria. He was too frail to continue the family business and went to university in Vienna instead to study mathematics, philosophy and astronomy. Before finding a university job in Prague, he had to work as a bookkeeper, and he even considered emigrating to America. Although promoted to professor, Doppler struggled with his teaching load, and his health suffered. One of his friends wrote: 'It is hard to believe how fruitful a genius Austria has in this man. I have written to . . . many people who can save Doppler for science and not let him die under the yoke. Unfortunately I fear the worst.' Doppler eventually left Prague and moved back to Vienna. In 1842, he presented a paper describing the colour shift in the light of stars that we now call the Doppler effect: 'It is almost to be accepted with certainty that this will in the not too distant future offer astronomers a welcome means to determine the movements and distances of such stars'. Although regarded as imaginative, he received a mixed reception from other prominent scientists. Doppler's detractors questioned his mathematical ability, whereas his friends thought very highly of his scientific creativity and intuition.

frequency drops. Light from the approaching planet is said to be 'blueshifted'; as it moves away it has a 'redshift'. Since the 1990s, hundreds of planets have been spotted around distant stars by finding this pattern imprinted in the glow of the central star.

Redshifts can arise not only due to planets' orbital motions, but also from the expansion of the universe itself, when it is called cosmological redshift. If the intervening space between us and a distant galaxy swells steadily as the universe expands, it is equivalent to the galaxy moving away from us with some speed. Similarly, two dots on a balloon being inflated look as if they are moving apart. Consequently the galaxy's light is shifted to lower frequencies because the waves must travel further and further to reach us. So very distant galaxies look redder than ones nearby. Strictly speaking, cosmological redshift is not a true Doppler effect because the receding galaxy is not actually moving relative to any other objects near it. The galaxy is fixed in its surroundings and it is the intervening space that is actually stretching.

To his credit, Doppler himself saw that the Doppler effect could be useful to astronomers, but even he could not have foreseen how much would flow from it. He claimed to have seen it recorded in the colours of light from paired stars, but this was disputed in his day. Doppler was an imaginative and creative scientist, but sometimes his enthusiasm outstripped his experimental skill. Decades later, however, galactic redshifts were measured by astronomer Vesto Slipher, setting the stage for the development of the Big Bang model of the universe. And now the Doppler effect may help identify worlds around distant stars that could even turn out to harbour life.

Unseen planet tugs distant star

Doppler Shift due to Stellar Wobble

Definition of redshift, z

Redshifts and blue shifts are expressed in terms of the proportional change in the observed and emitted wavelengths (or frequency) of an object. Astronomers refer to this scaling using the dimensionless symbol, z, such that the ratio of the observed to emitted wavelength equals 1 + z.

Redshifts, so defined, are used as shorthand for the distance to an astronomical object. For a galaxy with z = 1 for example, we observe its light at twice the wavelength at which it was emitted. Such an object would be about half way across the universe. The most distant galaxies known have z = 7 – 9, corresponding to around 80% of the universe. The cosmic microwave bakground, the most distant thing we can see, lies at z/approx1000.

the condensed idea
Stretched pitch

09 Parallax

How far away are the stars? The parallax method uses the fact that nearby objects appear to whiz by faster than more distant ones when seen from the moving Earth. The slight shift in the positions that results tells us that the closest stars are more than a million times further away than the Sun from Earth. Most are located within a disc that forms our own galaxy, which we see projected as a band on the sky that we call the Milky Way.

Once it began to be appreciated that the stars were not pinpricks in glass spheres but myriads of distant suns, the question arose of how far away they are. Their patterns were assigned names as constellations – such as Orion the Hunter, Ursa Major the Great Bear and Crux Australis the Southern Cross – but the question of how they are distributed in space has taken centuries to answer.

The first clue is that the stars are not spread uniformly on the sky, with the great majority lying in the pale band that we call the Milky Way. It is brightest in the southern hemisphere, especially near the constellation Sagittarius, where the timeless view is also pockmarked with black clouds and bright fuzzy patches called nebulae. We know today that the band of the Milky Way consists of billions of faint stars, blurred together by our eyes. If we map these positions in more detail, we see that the stars clump together in spiral arms; like soap suds swirling around a bathroom plughole, the Milky Way's stars spiral about the centre of our galaxy, pulled by gravity. The Sun is situated on one of these spiral arms, in a quiet galactic suburb. But how was this worked out?

The Milky Way Named in Latin as *Via Lactica*, The Milky Way intrigued the ancients. Greek philosophers including Aristotle and Anaxagoras

timeline

1573	1674
Digges proposed method of parallax	Hooke detected shift in position of γ Draconis

wondered if it was indeed a sea of distant burning stars. But they had no way of dissecting it. It was not until 1610, when Galileo applied his telescope, that the haze was broken up to reveal multitudes of individual stars.

The distribution of stars in space in three dimensions was pondered by philosopher Immanuel Kant. In a treatise published in 1755 he speculated that the Milky Way's stars lay in a giant disc held together by the force of gravity, just as the planets in our solar system orbit the Sun within a single plane. The stars form a band across the sky because we are viewing them from our location within that disc.

In 1785, British astronomer William Herschel measured the shape of the Milky Way disc in detail by painstakingly surveying hundreds of stars. Plotting their positions, he realised that there were many more of them in one part of the sky than in the opposite direction. He suggested that the Sun lay to one side of the Milky Way disc, not in the centre as had been previously supposed.

Far away Although the stars were once thought to all lie at about the same distance from the Earth, astronomers gradually realised that this was unlikely. Clearly they were unevenly scattered. Isaac Newton's theory of gravity implied that if they were massive they would be drawn toward one another, just as the planets are attracted to the Sun. But because all the stars were not in one clump, this attraction must be weak. Therefore the stars must lie very far apart. By this reasoning, Newton was one of the first to realize how distant the stars really were.

Astronomers sought methods for determining the distance to a star. One way was based on its brightness – if a star is as bright as the Sun, then its brightness should fade by the square of its distance. Using this

> ## Arc seconds
>
> Astronomers measure distances in the sky using projected angles. The Moon's size is about half a degree. Degrees are subdivided further into 60 arc minutes ('), which are broken into 60 arc seconds ("). So an arc second is 1/3600 of a degree.

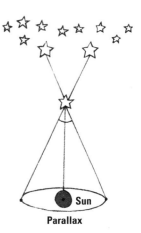

Parallax

1725	1755	1785	1838	1989
Bradley proposed theory of stellar aberration	Kant postulated that the Milky Way is a disc	Herschel measured disc shape of Milky Way	Bessel measured parallax	Hipparcos satellite launched

Parsecs

Measurements of stellar parallax are often defined as the difference in position of a star as seen from the Earth and Sun. This is equivalent to the angle subtended at a star by the mean radius of the Earth's orbit around the Sun. The parsec (3.26 light years) is defined as the distance for which this parallax is 1 arc second.

assumption, the Dutch physicist Christiaan Huygens (1629–95) worked out how far away the brightest star in the night sky, Sirius, was. By adjusting the size of a tiny hole in a screen, he was able to let through exactly the same amount of sunlight as the star. After working out the hole's size relative to that of the Sun, he concluded that Sirius must be tens of thousands of times further away. Newton later put Sirius's distance at a million times that of the Sun, by comparing the star's brightness with a planet. Newton was close – Sirius is around half that distance away. The vastness of interstellar space was revealed.

Parallax But all stars are not exactly as bright as the Sun. In 1573, British astronomer Thomas Digges proposed that the geographer's method of parallax might be applied to the stars. Parallax is a shift in the angle at which you view a landmark as you move by it; if you are travelling through a landscape, then the compass bearing to a nearby hilltop alters more quickly than the line to a mountain in the distance. Or, in a car, nearby trees whiz past quicker than ones further away. Nearby stars, viewed from the moving Earth as it follows its elliptical path around the Sun, should therefore move back and forth in the sky by some tiny amount each year, the amount depending on their distance from us.

Astronomers rushed to try to detect these annual shifts in the stars' positions, both to measure the distances to them and to confirm the heliocentric model of the solar system. Yet in doing so they found something else. In 1674, Robert Hooke published such an offset in the position of γ Draconis, a bright star that passes overhead at the latitude of London, allowing him to make accurate observations through a specially constructed hole in his roof. In 1680, Jean Picard reported that Polaris, or the Pole Star, also shifted its position by as much as 40 seconds of arc each year; and John Flamsteed, in 1689, confirmed it.

Curious as to what these measurements meant, James Bradley re-observed and confirmed the seasonal motion of γ Draconis in 1725 and 1726. But these shifts didn't look like parallax: stars should shift by different amounts depending on their distances, yet these were all shifting in the same way.

ROBERT HOOKE 1635–1703

Robert Hooke was born on the Isle of Wight in England, the son of a curate. He studied at Christ Church, Oxford, working as the assistant to physicist and chemist Robert Boyle. In 1660 he discovered Hooke's law of elasticity and soon after was appointed Curator of Experiments for meetings at the Royal Society. Publishing *Micrographia* five years later, Hooke coined the term 'cell', after comparing the appearance of plant cells under a microscope to the cells of monks. In 1666, Hooke helped rebuild London after the Great Fire, working with Christopher Wren on the Royal Greenwich Observatory, the Monument and Bethlem Royal Hospital (known as 'Bedlam'). He died in London in 1703 and was buried at Bishopsgate in London, but his remains were moved to north London in the 19th century and their current whereabouts are unknown. In February 2006 a long-lost copy of Hooke's notes from Royal Society meetings was discovered and is now housed at the Royal Society in London.

He was puzzled. A couple of years later, he realized what it was: just as a wind vane on a mast shifts when the boat changes its direction of motion, to show a combination of the directions of the wind and the boat, so the motion of the Earth was varying how we see the stars. The stars all nod slightly as we go round the Sun. This surprise discovery, called stellar aberration, also confirms that the Earth orbits the Sun.

> **'If I have seen further, it is by standing on the shoulders of giants.'**
>
> **Isaac Newton**

Parallax was not found until instruments became accurate enough. The first successful measurements were made by Friedrich Bessel in 1838 for the star 61 Cygni. Because the stars are so far away, the parallax they show is very small and hard to measure. For example, our nearest star, Proxima Centauri, has a parallax less than a second of arc, some 50 times smaller than its aberration. Today, satellites such as ESA's Hipparcos have measured accurate positions for 100,000 neighbourhood stars, allowing distances to be derived for many. Even so, parallaxes only reach across about one per cent of our galaxy.

the condensed idea
Foreground star shift

10 The Great Debate

A meeting of two minds in 1920 set the stage for the greatest change in man's thinking about the universe – the idea that our galaxy is but one of many that pepper space. Just as big a paradigm shift as the concept that the Earth goes round the Sun, and that the Sun is one among many stars, the Great Debate set out the questions to be tested to demonstrate that galaxies exist beyond the Milky Way.

How big is the universe? In 1920 this question boiled down to knowing the size of the Milky Way. Over the previous centuries, astronomers had come to terms with the idea that the stars were distant suns, similar to our own, and were spread across the sky in a flattened disc configuration. The plane of the disc projected on the sky formed the band of the Milky Way, which is also the name we give our galaxy.

But the Milky Way consists of more than just stars. It contains many fuzzy clouds, or 'nebulae', such as the smudge that lies in the belt of the constellation Orion, known as the Horsehead nebula because of the equestrian form of a striking dark cloud within it. Most of these nebulae are irregularly shaped, but a subset are elliptical with spiral patterns superimposed. A famous example is the Andromeda nebula, in the constellation of that name.

Other components of the Milky Way include clusters of stars, such as the Pleiades, a group of blue stars embedded in fuzz that is discernible with the naked eye. Denser star clusters also pepper the sky, including globular clusters, which are concentrated balls of hundreds of thousands of stars. Around 150 globular clusters are known in the Milky Way.

timeline

1665

Globular clusters discovered by German amateur astronomer Abraham Ihle

1784

Cepheid variable stars discovered

At the beginning of the twentieth century, astronomers were starting to map the geometry of the heavens by piecing together the distributions of these objects in three-dimensional space. They sought in particular the detailed shape of the Milky Way, which was then assumed to contain everything in the known universe.

The debate On 26 April 1920, two great American astronomers went head to head to debate the size of the Milky Way. They met at the Smithsonian Museum of Natural History in Washington DC, following a meeting of the US's National Academy of Sciences. In the audience were many top scientists, allegedly including Albert Einstein. The debate is credited with setting out the logic that would precipitate a change in our understanding of the scale of the universe.

First to speak was Harlow Shapley, a bright young astronomer from Mount Wilson Observatory in California. He faced the more established figure of Heber Curtis, director of the Allegheny Observatory in Pittsburgh, Pennsylvania. The two presented their arguments about the size of the Milky Way, based on the different astronomical yardsticks in which they were expert.

Andromeda galaxy

Shapley had measured the distances to globular clusters. He found they were much further away than he had anticipated, implying that our galaxy was 10 times bigger than had been thought – it was some 300,000 light years in diameter. He also saw that there were more globular clusters in one half of the sky than the other, indicating that the Sun lay far from the centre – he estimated it was 60,000 light years out, or about halfway. Such a picture was shocking. The Sun was an average star, nowhere near the centre of things.

1789	**1908**	**1920**	**1924**
Herschel catalogues and names globular clusters	Henrietta Swan Leavitt discovers Cepheid properties that indicate distance	Shapley v. Curtis Great Debate	Hubble measures distance to Andromeda nebula well beyond Milky Way

Curtis, meanwhile, was focused on understanding a different problem – the nature of the spiral nebulae. The peculiar characteristics of these structured clouds led him and others to believe that they were a distinct class of objects lying beyond the boundaries of the Milky Way. This belief fitted with the then assumed small radius of the Milky Way.

The clash between the results of the two astronomers suggested a major problem that needed to be solved. Shapley's new measurements had extended the Milky Way by such a large amount that the possibility that Curtis's nebulae lived outside it was called into question. Nevertheless, the peculiar nebulae seemed unlike anything else within the Milky Way. A closer look at the evidence was needed.

The arguments Both astronomers presented data to back up their ideas. Shapley stood by his globular cluster distance measurements, concluding that the Milky Way was so large that everything we see in the night sky must be contained in it. His technique used a particular type of variable star whose flashing period gives away its brightness – it is called a Cepheid variable star, after its prototype Delta Cephei. Essentially these bright pulsating stars act like light bulbs of known wattage, so their distance can be ascertained.

Curtis was more cautious. He countered that the Milky Way couldn't be so large – perhaps the Cepheid distances weren't correct – and the properties of the spiral nebulae were such that they must lie outside it. The spiral nebulae behaved like miniature versions of our own galaxy. Like the Milky Way, they contained exploding stars in similar numbers, they rotated in a similar way to our own, they were about the same size and some had dark

Light years

A light year is the distance that light travels in one year. Light moves at a velocity of about 300,000 km each second. So in one year, it travels about 10 trillion km. The Milky Way is about 150,000 light years across; the Andromeda galaxy is 2.3 million light years away.

Astronomical units

In our solar system, astronomers sometimes use a unit of distance called the Astronomical Unit (AU). The AU is defined as the average distance between the Earth and the Sun and is about 150 million km (93 million miles). Mercury is about 1/3 of an AU from the Sun and Pluto averages about 40 AU from the Sun.

lanes across their longest axis, suggesting they were disc-like. They looked as if they were other galaxies, implying that ours was not the only one.

Who was right? The debate was a draw; there was no clear winner. Both were right in part, both wrong in some ways. Each was correct about his own speciality. Shapley's distances were about right. And the Sun does lie off centre. But more importantly, Curtis was fundamentally right that the nebulae lie beyond our galaxy – they are 'island universes'. The proof came in 1924, when Edwin Hubble combined both sets of evidence. He measured the distance to the Andromeda nebula – one of our nearest neighbour galaxies – using Shapley's technique with Cepheid variable stars, and found that it was much further away than the globular clusters. It was indeed well beyond the Milky Way.

Implications Although the debate was more an airing of arguments than a sparring match with a clear victory, it set out the questions that astronomers needed to test. It was thus a focal point for the transformation of our thinking about the scale of the universe.

Just as Copernicus knocked the Earth away from the universe's heart in favour of the Sun, Shapley knocked the Sun off centre in favour of the core of the Milky Way. Curtis went even further and showed that the Milky Way isn't unique and special – it is but one of billions of other galaxies. Mankind's place in the universe really is precarious.

the condensed idea
Realm of the Galaxies

11 Olbers' paradox

Why is the night sky dark? If the universe were endless and had existed for ever, then it should be as bright as the Sun, yet it is not. Looking up at the night sky you are viewing the entire history of the universe. The number of stars is limited and implies that the universe has a limited finite and age. Olbers' paradox paved the way for modern cosmology and the Big Bang model.

You might think that mapping the entire universe and viewing its history would be difficult and call for expensive satellites in space, huge telescopes on remote mountaintops, or a brain like Einstein's. But if you go out on a clear night you can make an observation that is every bit as profound as general relativity. The night sky is dark. Although this is something we take for granted, the fact that it is dark and not as bright as the Sun tells us a lot about our universe.

Star light star bright If the universe were infinitely big, extending for ever in all directions, then in every direction we looked we would eventually see a star. Every sight line would end on a star's surface. Going further away from the Earth, more and more stars would fill space. It is like looking through a forest of trees – nearby you can distinguish individual trunks, appearing larger the closer they are, but more and more distant trees fill your view. So if the forest was really big, you would not be able to see the landscape beyond. This is what would happen if the universe were infinitely big. Even though the stars are more widely spaced than the trees, eventually there would be enough of them to block the entire view. If all the stars were like the Sun, then every point of sky would be filled with starlight. Even though a single star far away is faint, there are more stars at

timeline

1610

Kepler notes that the night sky is dark

that distance. If you add up all the light from those stars, they provide as much light as the Sun, so the entire night sky should be as bright as the Sun.

Obviously this is not so. The paradox of the dark night sky was noted by Johannes Kepler in the 17th century, but only formulated in 1823 by German astronomer Heinrich Olbers. The solutions to the paradox are profound. There are several explanations, and each one has elements of truth that are now understood and adopted by modern astronomers. Nevertheless, it is amazing that such a simple observation can tell us so much.

End in sight The first explanation is that the universe is not infinitely big. It must stop somewhere. So there must be a limited number of stars in it and not all sight lines will find a star. Similarly, standing near the edge of the forest or in a small wood, you can see the sky beyond.

Another explanation could be that the more distant stars are fewer in number, so they do not add together to give as much light. Because light travels at a precise speed, the light from distant stars takes longer to reach us than that from nearby stars. It takes eight minutes for light to reach us from the Sun but four years for light from the next nearest star, Alpha Centauri, to arrive, and as much as 100,000 years for light to get to us from stars on the other side of our own galaxy. Light from the next nearest galaxy, Andromeda, takes two million years to reach us; it is the most distant object we can see with the naked eye. So as we peer further into the universe, we are looking back in time, and distant stars appear younger than the ones nearby because their light has travelled a long time to reach us. This could help us with Olbers' paradox if those youthful stars are rarer

Dark skies

The beauty of the dark night sky is becoming harder to see due to the glow of lights from our cities. On clear nights throughout history people have been able to look upward and see a brightly lit backbone of stars, stretched across the heavens. Even in large cities 50 years ago it was possible to see the brightest stars and the Milky Way's swath, but nowadays hardly any stars are visible from towns and even the countryside views of the heavens are washed out by yellow smog. The vista that has inspired generations before us is becoming obscured. Sodium streetlights are the main culprit, especially ones that waste light by shining upwards as well as down. Groups worldwide, such as the International Dark-Sky Association, which includes astronomers, are now campaigning for curbs on light pollution so that our view out to the universe is preserved.

than stars nearby. Stars like the Sun live for about 10 billion years (bigger ones live for shorter times and smaller ones for longer), so the fact that stars have a finite lifetime could also explain the paradox. Stars do not exist in the very early universe because there is no time for them to have been born. So stars have not existed for ever.

Distant stars may also be fainter than the Sun because of redshift. The expansion of the universe stretches light wavelengths, causing the light from distant stars to appear redder. So stars a long way away will look a little cooler than stars nearby. This could also restrict the amount of light reaching us from the outermost parts of the universe.

Wackier ideas have also been put forward, such as the distant light being blocked out by soot from alien civilizations, iron needles or weird grey dust. But any absorbed light would be re-radiated as heat and so would turn up elsewhere in the spectrum. Astronomers have checked the light in the night sky at all wavelengths, from radio waves to gamma rays, and they have seen no sign that the visible starlight is blocked.

> **Were the succession of stars endless, then the background of the sky would present us an uniform luminosity, like that displayed by the Galaxy – since there could be absolutely no point, in all that background, at which would not exist a star.**
> **Edgar Allan Poe**

Middle-of-the-road universe So, the simple observation that the night sky is dark tells us that the universe is not infinite. It has only existed for a limited amount of time, it is restricted in size, and the stars in it have not existed for ever.

Modern cosmology is based on these ideas. The oldest stars we see are around 13 billion years old, so we know the universe must have been formed before this time. Olbers' paradox suggests it cannot be very much ahead of this or we would expect to see many previous generations of stars, and we do not.

Distant galaxies of stars are indeed redder than nearby ones, due to redshift, making them harder to see with optical telescopes and confirming that the universe is expanding. The most distant galaxies known today are so red they have become invisible and can only be picked up at infrared wavelengths.

Astronomers have dubbed the period during which the first stars switched on, and where galaxies are so reddened that they all but disappear from view, the cosmic 'dark ages'. It is a goal to try to find these first objects, to try to understand what made them form in the first place and how stars and galaxies grow from tiny seeds under gravity.

In postulating his paradox, Olbers didn't know it, but he was asking the very questions that absorb cosmologists today. So all this evidence supports the idea of the Big Bang, the theory that the universe grew out of a vast explosion some 14 billion years ago.

the condensed idea
Our finite universe

12 Hubble's law

American astronomer Edwin Hubble was the first to realize that galaxies beyond our own are all moving away from us together. The further away they are, the faster they recede, following Hubble's law. This galactic diaspora formed the first evidence that the universe is expanding, an astounding finding that changed our view of our entire universe and its destiny.

Copernicus's deduction in the 16th century that the Earth goes around the Sun caused major consternation. Humans no longer inhabited the exact centre of the cosmos. But in the 1920s, Hubble made telescope measurements that were even more unsettling. He showed that the entire universe was not static but expanding.

Hubble mapped out the distances to other galaxies and their relative speeds compared with our Milky Way, and found that they were all hurtling away from us. We were so cosmically unpopular that only a few close neighbours were inching towards us. The more distant the galaxy, the faster it was receding, with a speed proportional to its distance away (Hubble's law). The ratio of speed to distance is always the same number and is called the Hubble constant. Astronomers today have measured its value to be close to 72 km per second per megaparsec (a megaparsec, or a million parsecs, is equivalent to 3,262,000 light years or 3×10^{22} m). Galaxies continually recede from us by this amount.

> **❛The history of astronomy is a history of receding horizons.❜**
> **Edwin Hubble**

timeline

1918	1920
Vesto Slipher measures redshifts of nebulae	Shapley and Curtis debate the size of the Milky Way

The Great Debate Before the twentieth century, astronomers barely understood our own galaxy, the Milky Way. They had measured hundreds of stars within it but also noted it was marked with many faint smudges, called nebulae. Some of these nebulae were gas clouds associated with the births and deaths of stars. But some looked different. Some had spiral or oval shapes that suggested they were more regular than a cloud. The origin of these nebulae was debated in 1920 by two famous astronomers (see p.41). Harlow Shapley argued that everything in the sky was part of the Milky Way; Heber Curtis proposed that some of these nebulae were external

... wed that the spiral nebulae ... iddenly opened up into a

... Mount Wilson to measure ... nebula, now known to be a ... lso a sibling in the group of ... are called Cepheid ... s of distance. The amount ... brightness of the star, so ... w bright it is. Knowing its ... is, because it is dimmed ... easured the distance to the ... n the size of the Milky ... l the debate –

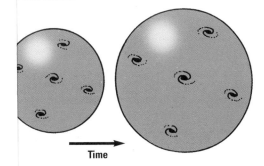

Time

1922
Alexander Friedman
publishes the Big
Bang model

1929
Hubble and Milton
Humason discover
Hubble's law

2001
Hubble Space Telescope
publishes accurate Hubble
constant value

than expected meant that these galaxies were rushing away from us, like many ambulance sirens falling off in tone as they speed away. It was very strange that all the galaxies were rushing away, with only a few close ones moving towards us. Moreover, the further away you looked, the faster they receded.

Hubble also saw that the galaxies weren't simply receding from us, which would have made our place in the universe very privileged. They were all hurtling away from each other too. He concluded that the universe itself was expanding, being inflated like a giant balloon. The galaxies are like spots on the balloon, getting further apart from one another as more air is added.

How far how fast? Even today astronomers use Cepheid variable stars to map out the local universe's expansion. Measuring the Hubble constant accurately has been a major goal. To do so you need to know how far away something is and its speed or redshift. Redshifts are straightforward to measure from spectral lines. The frequency of a particular atomic transition in starlight can be checked against its known wavelength in the laboratory; the

Hubble Space Telescope

The Hubble Space Telescope is surely the most popular satellite observatory ever. Its stunning photographs of nebulae, distant galaxies and discs around stars have graced the front pages of newspapers for 20 years. Launched in 1990 from the space shuttle Discovery, the spacecraft is about the size of a double-decker bus, 13 m long, 4 m across and weighing 11,000 kg. It carries an astronomical telescope whose mirror is 2.4 m across and a suite of cameras and electronic detectors that are able to take crystal-clear images, in visible and ultraviolet light and in infrared. Hubble's power lies in the fact that it is located above the atmosphere – so its photographs are not blurred. Now getting old, its fate is uncertain. Its instruments having been upgraded for the last time, when NASA terminates its programme it may either rescue the craft for posterity or crash it safely into the ocean.

> **❝We find them smaller and fainter, in constantly increasing numbers, and we know that we are reaching into space, farther and farther, until, with the faintest nebulae that can be detected with the greatest telescopes, we arrive at the frontier of the known universe.❞**
>
> **Edwin Hubble**

difference gives its redshift. Distances are harder to determine, because you need to observe something far away in the galaxy either whose true distance is known or whose true brightness can be measured: a 'standard candle'.

There are a variety of methods for inferring astronomical distances. Cepheid stars work for nearby galaxies when you can separate the individual stars. But further away, other techniques are needed. All the different techniques can be tied together one by one to build up a giant measuring rod, or 'distance ladder'. But because each method comes with peculiarities, there are still many uncertainties in the accuracy of the extended ladder.

The Hubble constant is now known to an accuracy of about 10 per cent, thanks largely to observations of galaxies with the Hubble Space Telescope and of the cosmic microwave background radiation. The expansion of the universe began in the Big Bang, the explosion that created the universe, and galaxies have been flying apart ever since then. Hubble's law sets a limit on the age of the universe. Because it is continuously expanding, if you trace the expansion back to the beginning point, you can work out how long ago that was. It turns out to be around 14 billion years. This expansion rate is fortunately not enough to break apart the universe. The cosmos instead is finely balanced, in between completely blowing apart and containing enough mass to collapse back in on itself eventually.

the condensed idea
The expanding universe

13 Cosmic distance ladder

Different measures of astronomical distance have led to great paradigm shifts in astronomy. Distances to the stars left us feeling small; determining the size of the Milky Way and the remoteness of nearby nebulae opened up the cosmos of galaxies. Because the scales are so vast, no single method works across the entire universe. The cosmic distance ladder is the patchwork that results from bolting together a series of techniques.

Because the universe is so big, measuring distances right across the universe is challenging. A rod that works within our galaxy cannot stretch to the furthest reaches of the cosmos. So a raft of different methods has been developed, each technique applied to a different range. Where methods overlap, the adjacent scales can be tied together, building up a series of steps that is known as the 'cosmic distance ladder'. The rungs of the ladder step out across the universe, from the solar neighbourhood to the nearest stars, across the Milky Way to other galaxies, galaxy clusters and the edge of the visible universe.

The first rung is the firmest. Nearby stars can be accurately positioned using the trigonometric method of parallax. Just as a hiker locates a far mountaintop on his map by taking a series of bearings as he walks along, so an astronomer sited on the moving Earth can orient a star by measuring its shift in position against the more distant background stars. The degree of shift tells the astronomer how far away the star is: those nearby move more than those far away. But the distance to the stars is so great – the nearest

timeline

1784	**1918**
Cepheid variable stars discovered	Cepheid distance scale worked out

star is four light years away – that the shifts are tiny and hard to measure. Parallaxes can only cover a fraction of the Milky Way. To go further, new methods are needed.

Cepheids Unique stars form the next step. If you know exactly how bright a star is – as if it was the cosmic equivalent of a scaled-up 100-watt light bulb, known as a 'standard candle'– then you can work out how far away it is by measuring its faintness. Brightness falls off with the square of distance; so a star twice as far away as another identical one will appear four times weaker. But the trick is to know the intrinsic brightness of that star. Stars come in all shapes, sizes and colours – from red giants to white dwarfs – so this isn't straightforward. For rare types of stars, there is a way.

Cepheid variable stars are very useful standard candles. The wattage of the star is given away by the rate at which it flickers. Compare that with how faint the star appears on the sky and you know how far away it is. Cepheid-type stars are bright enough to be seen across the whole Milky Way and even out in other galaxies beyond our own. So they can be used to scope out the local region of the universe around our galaxy.

Cosmic dust

One problem with using standard candles at large distances is that they may be dimmed by intervening material. Galaxies are messy places, full of gas clouds, debris and carbon-rich soot, and should your star or supernova fall behind some polluting smog then it might appear dimmer than it really is. Astronomers try to get around this by looking carefully for indicators of cosmic soot. One obvious sign is that it changes the colour of the background star, making it appear redder, as when dramatic sunsets followed the injection of dust into Earth's atmosphere by the 1991 Mount Pinatubo volcanic eruption. If astronomers spot the signs of dust, they can correct the star's brightness accordingly.

1924
Hubble measures distance
to Andromeda galaxy

1929
Hubble measures
cosmic expansion

1998
Supernova data
indicate dark energy

Supernovae Stepping out further, even brighter standard candles are needed. Amongst stars, the most powerful lighthouses are supernovae, the catastrophic explosions of dying suns. One particular class, called a Type Ia supernova, is extremely valuable and can be detected out to quite distant reaches of the universe. The exact brightness of a Type Ia supernova can be determined from the rate at which it explodes, first flaring up and then fading.

Supernova happen rarely – one may go off every 50 years in a galaxy the size of the Milky Way – so they are most useful at cosmic distances where there are many galaxies available to increase the chances of seeing one during your astronomy career. Supernovae in distant galaxies have indicated that the expansion of the universe is affected by a mysterious component called dark energy – a sort of anti-gravity term in the equations of general relativity (see p.92).

Redshift At cosmic scales, redshifts of spectral lines are the most widely used distance indicators. According to Hubble's law, the more distant a galaxy is, the faster it recedes from us due to the expansion of the universe, and the more shifted to the red end of the spectrum are its chemical emission and absorption lines. But because it only indicates the gross speed of a galaxy, a redshift may be contaminated by the object's local motions. Redshifts are therefore good as a rough indicator, but less useful for accurate determinations of distance and less useful nearby when intrinsic motions might be similar in scale to universal expansion speeds. Today, galaxies are visible across about 80 per cent of the universe. Astronomers compete each year to better this record.

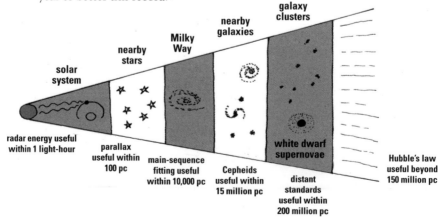

> **The rung of a ladder was never meant to rest upon, but only to hold a man's foot long enough to enable him to put the other somewhat higher.**
> **Thomas Huxley**

Statistical methods A range of other methods have been tried. Some are geometric, comparing 'rulers' whose actual length can be determined by applying basic physics theories to scales measured on the sky. These include average distances between clusters of galaxies, and characteristic sizes of hot and cold patches in the cosmic microwave background.

Statistical methods also work. Because the life cycles of stars are well understood, certain phases can be used as indicators. Just as individual Cepheids give away distances by their brightness and period, averaged statistics can pinpoint key changes in the brightness and colour of populations of thousands of stars. Another technique that is used for galaxies is akin to determining distance from how blurred a galaxy appears – a galaxy made up of billions of stars appears grainy when seen close up, and smoother from afar as individual stars are blurred out.

The cosmic distance ladder has a firm footing, but gets a bit more rickety as it steps out into space. Even so, the vastness of space means that that doesn't matter too much. From the nearest stars a few light years away, to the edge of our Milky Way 100,000 light years across, distances are well measured. The cosmic expansion kicks in beyond our local group of galaxies, more than 10 million light years away, and so distances become harder to interpret. Yet standard candles have revealed not only that our universe is expanding but that dark energy exists, and they have tied everything in to the fundamental physics of the early universe. Perhaps it's not so rickety after all.

the condensed idea
Patchwork of scales

14 The Big Bang

The birth of the universe occurred in a phenomenal explosion that created all space, matter and time as we know it. Predicted from the mathematics of general relativity, we see evidence for the Big Bang in the rush of galaxies away from our own, the quantities of light elements in the universe and the microwave glow that fills the sky.

The Big Bang is the ultimate explosion – the birth of the universe. Looking around us today, we see signs that our universe is expanding and infer that it must have been smaller, and hotter, in the past. Taking this to its logical conclusion means that the entire cosmos could have originated from a single point. At the moment of ignition, space and time and matter were all created together in a cosmic fireball. Very gradually, over 14 billion years, this hot, dense cloud swelled and cooled. Eventually it fragmented to produce the stars and galaxies that dot the heavens today.

It's no joke The 'Big Bang' phrase itself was actually coined in ridicule. The eminent British astronomer Fred Hoyle thought it preposterous that the whole universe grew from a single seed. In a series of lectures first broadcast in 1949, he derided as far-fetched the proposition of Belgian mathematician Georges Lemaître, who found such a solution in Einstein's equations of general relativity. Instead, Hoyle preferred to believe in a more sustainable vision of the cosmos. In his perpetual 'steady state' universe, matter and space were being continually created and destroyed and so could have existed for an unlimited time. Even so, clues were already amassing, and by the 1960s, Hoyle's static picture had to give way, due to the weight of evidence that favoured the Big Bang.

timeline

1927	**1929**
Friedmann and Lemaître devise Big Bang theory	Hubble detects the expansion of the universe

The expanding universe Three critical observations underpin the success of the Big Bang model. The first is Edwin Hubble's observation in the 1920s that most galaxies are moving away from our own. Looked at from afar, all galaxies tend to fly apart from one another as if the fabric of space–time is expanding and stretching, following Hubble's law. One consequence of the stretching is that light takes slightly longer to reach us when travelling across an expanding universe than one where distances are fixed. This effect is recorded as a shift in the frequency of the light, called the 'redshift' because the received light appears redder than it was when it left the distant star or galaxy. Redshifts can be used to infer astronomical distances.

> **There is a coherent plan in the universe, though I don't know what it's a plan for.**
>
> **Fred Hoyle**

Light elements Going back in time to the first hours of the newborn universe, just after the Big Bang, we have to imagine everything packed close together in a seething superheated cauldron. In the first seconds, the universe was so hot and dense that not even atoms were stable. As it grew and cooled, a particle soup emerged, stocked with quarks, gluons and other fundamental particles. After just a minute the quarks stuck together to form protons and neutrons. Then, within the first three minutes, cosmic chemistry mixed the protons and neutrons, according to their relative numbers, into atomic nuclei. This is when elements other than hydrogen were first formed by nuclear fusion. Once the universe had cooled below the fusion limit, no elements heavier than beryllium could be made. So the universe was initially awash with the nuclei of hydrogen and helium and traces of deuterium (heavy hydrogen), lithium and beryllium created in the Big Bang itself.

In the 1940s, Ralph Alpher and George Gamow predicted the proportions of light elements produced in the Big Bang, and this basic picture has been confirmed by even the most recent measurements of slow-burning stars and primitive gas clouds in our Milky Way.

Microwave glow Another pillar supporting the Big Bang theory is the discovery in 1965 of the faint echo of the Big Bang itself. Arno Penzias and

1948	**1949**	**1965**	**1992**
The cosmic microwave background is predicted Big Bang nucleosynthesis is calculated by Alpher and Gamow	Hoyle coins the term 'Big Bang'	Penzias and Wilson detect the cosmic microwave background	COBE satellite measures cosmic microwave background patches

Robert Wilson were working on a radio receiver at Bell Labs in New Jersey when they were puzzled by a weak noise signal they could not get rid of. It seemed there was an extra source of microwaves coming from all over the sky, equivalent to something a few degrees in temperature. They had stumbled upon the cosmic microwave background radiation, a sea of photons left over from the very young hot universe.

In Big Bang theory, the existence of the microwave background had been predicted in 1948 by George Gamow, Ralph Alpher and Robert Hermann. Although nuclei were synthesized within the first three minutes, atoms were not formed for 400,000 years. Eventually, negatively charged electrons paired with positively charged nuclei to make atoms of hydrogen and light elements. The removal of charged particles, which scatter and block the path of light, cleared the fog and made the universe transparent. From then onwards, light could travel freely across the universe, allowing us to see back that far.

Although the young universe fog was originally hot (some 3,000 kelvins or K), the expansion of the universe has redshifted the glow from it so that we see it today at a temperature of less than 3 K (three degrees above absolute zero). This is what Penzias and Wilson spotted. So with these three major foundations so far intact, Big Bang theory is widely accepted by most astrophysicists. A handful still pursue the steady state theory that attracted Fred Hoyle, but it is difficult to explain all these observations in any other model.

Fate and past What happened before the Big Bang? Because space–time was created in the Big Bang, this is not really a very meaningful question to ask – a bit like 'Where does the Earth begin?' or 'What is north of the north pole?' However, mathematical physicists do ponder the triggering of the Big Bang in multi-dimensional space (often 11 dimensions) through the mathematics of M-theory and string theory. These look at the physics and energies of strings and membranes in these multi-dimensions and incorporate ideas of particle physics and quantum mechanics to try to understand how such an event was triggered. With parallels with quantum physics ideas, some cosmologists also discuss the existence of parallel universes.

In the Big Bang model, unlike the steady state model, the universe evolves. The cosmos's fate is dictated largely by the balance between the amount of

Big Bang timeline

13.7 billion years (after the Big Bang): now (temperature T = 2.726 K)

200 million years: 'reionization'; first stars heat and ionize hydrogen gas (T = 50 K)

380,000 years: 'recombination'; hydrogen gas cools down to form molecules (T = 3,000 K)

10,000 years: end of the radiation-dominated era (T = 12,000 K)

1,000 seconds: decay of lone neutrons (T = 500 million K)

180 seconds: 'nucleosynthesis'; formation of helium and other elements from hydrogen (T = 1 billion K)

10 seconds: annihilation of electron–positron pairs (T = 5 billion K)

1 second: decoupling of neutrinos (T = 10 billion K)

100 microseconds: annihilation of pions T = 1 trillion K)

50 microseconds: 'QCD phase transition'; quarks bound into neutrons and protons (T = 2 trillion K)

10 picoseconds: 'electroweak phase transition'; electromagnetic and weak force become different (T = 1–2 quadrillion K)

Before this time the temperatures were so high that our knowledge of physics is uncertain.

matter pulling it together through gravity and other physical forces that pull it apart, including the expansion of the universe. If gravity wins, then the universe's expansion could one day stall and it could start to fall back in on itself, ending in a rewind of the Big Bang, known as the big crunch. Universes could follow many of these birth–death cycles. Alternatively, if the expansion and other repelling forces (such as dark energy) win, they will eventually pull all the stars and galaxies and planets apart and our universe could end up a dark desert of black holes and particles, a 'big chill'. Lastly there is the 'Goldilocks universe', where the attractive and repellent forces balance and the universe continues to expand for ever but at a gradually slower rate. It is this ending that modern cosmology is pointing to as being most likely. Our universe is just right.

the condensed idea
The ultimate explosion

15 Cosmic micro- wave background

The discovery of the cosmic microwave background cemented the Big Bang theory. Originating in the heat of the very early universe, this sea of weak electromagnetic radiation is due to photons that were released over 13 billion years ago when space became transparent at the time when hydrogen atoms formed.

In 1965, Arno Penzias and Robert Wilson discovered an unexpected warm glow in the sky. Working on their microwave radio antenna in New Jersey, the Bell Labs physicists found a faint heat signal emanating from every direction that wouldn't go away. At first they assumed it was mundane – perhaps due to pigeon droppings clogging their sensitive horn.

But after hearing a talk by Princeton theorist Robert Dicke, they realized they had stumbled on a huge discovery. The bath of warmth that they had seen wasn't coming from the Earth; its origin was cosmic. They had found the predicted afterglow of the Big Bang. Dicke, who had built a similar radio antenna to look for the background radiation, was a little less jubilant: 'Boys, we've been scooped,' he quipped.

Warm glow The cosmic microwave background makes the whole sky appear as a bath of warmth with a temperature of about 3 degrees Kelvin (equivalent to 3 degrees Celsius above absolute zero). Its characteristics are precisely predicted by the physics of the Big Bang. When the universe was young, it was scorching hot, reaching thousands of degrees K. But as it expanded, it cooled down. Today it should be exactly 2.73 K; and that is what Penzias and Wilson found.

timeline

1901	1948
Max Planck explains black-body radiation using quanta	Ralph Alpher and Robert Herman predict a 5 K cosmic background in their theory

The cosmic microwave background has the most well-defined temperature of any source. No man-made device in any laboratory has done better. The sky emits microwaves in a frequency range that peaks around 160.2 GHz (1.9 mm wavelength), and is a perfect example of a 'black-body spectrum' – a characteristic frequency range given off by something that absorbs and emits heat perfectly, such as a matt-black stove. In 1990, NASA's Cosmic Background Explorer (COBE) satellite showed that the cosmic microwave background is the most perfect example of a black-body spectrum ever seen, albeit much colder than a red-hot poker.

Dipole If you look carefully, the sky is not exactly the same temperature all over. The microwaves appear warmer in one hemisphere by 2.5 milli-Kelvin, or about one part in a thousand. Discovered soon after the background radiation itself, this heat pattern is known as the 'dipole', due to its two poles, hot and cold. This temperature difference arises from the Doppler effect, due to the motion of the Earth: the solar system is moving at 600 km/s relative to the universe.

If you look closer again, at a level of about one part in a million, the sky is speckled with hot and cold spots. These ripples are of great interest to astronomers because they were imprinted shortly after the Big Bang. They were first seen in 1992 by NASA's COBE satellite, which revealed numerous patches about the size of the full Moon. In 2003, a more detailed map was revealed by the Wilkinson Microwave Anisotropy Probe (WMAP) satellite, which broke the spots up into a further rash. And they will be measured in further exquisite detail by another satellite called Planck.

> **Change is rarely comfortable.**
> **Arno Penzias**

WMAP

COBE

1965	1990	1992	2009
Penzias and Wilson observe the cosmic microwave background	NASA's COBE satellite accurately measures the temperature of the microwave background	NASA's COBE discovers cosmic ripples	ESA's Planck satellite launched

Ripples These ripples in the cosmic microwave background originated back when the universe was extremely hot. After the Big Bang, the cosmos expanded and cooled and photons, subatomic particles and eventually protons and electrons formed. The nuclei of the first light elements, including hydrogen and a little helium and lithium, were created within three minutes. At this point the universe was a soup of protons and electrons, whizzing around. These particles had an electric charge – they were ionized: protons positive, electrons negative – but photons ricocheted off the charged particles, so the very early universe was an opaque fog.

The universe cooled further. The protons and electrons began to move more slowly, and after about 400,000 years they were eventually able to stick together, to form hydrogen atoms. Over this period the charged particles were gradually combined, and the nature of the cosmic soup changed, from being ionized to being electrically neutral. The universe became a sea of hydrogen.

Once the charged particles were mopped up, photons could travel freely. Suddenly we can see. These very photons, cooled even further, are the ones

Black-body radiation

Barbecue coals and electric stove rings turn red, orange and then yellow as they heat up, reaching hundreds of degrees Celsius. A tungsten light bulb filament glows white as it reaches over 3,000 degrees Celsius, similar to the surface of a star. With increasing temperature, hot bodies glow first red, then yellow and eventually blue-white. This spread of colours is described as black-body radiation because dark materials are best able to radiate or absorb heat. 19th century physicists found it hard to explain why this pattern held, irrespective of the substance they tested. Wilhelm Wien,

Lord Rayleigh and James Jeans worked out partial solutions. But Rayleigh and Jeans's solution was problematic in that it predicted that an infinite amount of energy would be released at ultraviolet wavelengths and above – the 'ultraviolet catastrophe'. Max Planck solved the problem in 1901 by joining the physics of heat and light together, dividing up the electromagnetic energy among a set of tiny subatomic units of electromagnetic field called 'quanta'. Planck's idea planted a seed that would grow to become one of the most important areas of modern physics: quantum theory.

> **'Scientific discovery and scientific knowledge have been achieved only by those who have gone in pursuit of it without any practical purpose whatsoever in view.'**
>
> **Max Planck**

that now make up the cosmic microwave background. At this time, corresponding to a redshift of around a thousand ($z = 1,000$), the universe had a temperature of roughly 3,000 K; it is now about 1,000 times cooler, around 3 K.

Cosmic landscape The hot and cold spots that fleck this bath of photons arise because of the matter in the universe. Some regions of space contained more matter than others, so that the photons travelling through them were slowed by slightly different amounts, depending on their path. The precise pattern of the microwave ripples tells us a lot about how unevenly matter was distributed well before any stars or galaxies had formed.

The typical scale of the hotspots is also telling. The most common size is about one degree on the sky, twice the full Moon's diameter. This is exactly what theorists have predicted by looking at the pattern of matter in the universe today and projecting it backwards, taking into account the universe's expansion. This close match between predicted and observed scales implies that light rays must travel in straight lines right across the universe: astronomers say that the universe is 'flat', because the rays do not bend or curve due to distortions in spacetime.

Overall, the story of the cosmic microwave background has been one of triumph for theorists. So far they have predicted its characteristics almost to the letter. But there is a chance that observers will find discrepancies suggestive of new physics, either in hotspot data from the Planck satellite, or in polarized signatures coming from experiments underway at the south pole, on balloons and using specialist radio telescopes.

the condensed idea
The universe's warm bath of photons

16 Big Bang nucleosynthesis

The lightest elements were created in the first minutes of the hot young universe in proportions that confirm the predictions of Big Bang theory. The amounts of helium, lithium and deuterium seen today in pristine regions of space are roughly as expected from that theory, which at the same time explains why these elements are surprisingly common in stars. That the level of deuterium is set low, however, implies that the universe is filled with exotic forms of matter.

A critical observation that backs up the Big Bang theory is the abundance of light elements in the universe. Nuclear reactions in the hot fireball phase of the Big Bang cooked up the first few atomic nuclei in precise ratios. Heavier ones were formed later from these initial ingredients by burning in the cores of stars.

Hydrogen is the most common element in the universe, and the major by-product of the Big Bang. Hydrogen is also the simplest element: a single proton orbited by an electron. It is sometimes found in a heavier form called deuterium, which consists of a normal hydrogen atom with an extra neutron, making it twice as heavy; a rarer form is tritium, which has a second neutron thrown in. The next element is helium, comprised of two protons, two neutrons and two electrons; then comes lithium with three protons, usually four neutrons and three electrons. These were all created in the early universe in a process called nucleosynthesis.

timeline

1920

Arthur Eddington suggests that fusion drives the stars

Alpher Bethe gamow paper

The theory of Big Bang nucleosynthesis was published in a 1948 paper
with a whimsical touch. Although its basics were initially worked out by
Ralph Alpher and George Gamow, due to the similarity of their surnames
to the first three Greek letters (alpha, beta, gamma), they asked Hans
Bethe to join them. The paper still amuses in physics circles.

Cooking with gas Just after the Big Bang explosion, the universe was
so hot that it was a boiling soup of fundamental particles. As it expanded
and cooled down, different particles came into being, eventually producing
the familiar protons, neutrons and electrons that make up the objects in our
world. When the universe was just three minutes old, its billion-degree
temperature was suitable for the nuclei of the lightest elements to be
created. Protons and neutrons were able to collide and stick together to
form deuterium, whose nuclei could combine further to produce helium.
Small amounts of tritium were possible, and a little lithium was made by
joining tritium with two deuterium nuclei.

Assuming that a certain number of protons and
neutrons were available in the hot young universe
as ingredients for this cosmic cookery, the relative
amounts of each light element can be predicted from
the nuclear reaction recipes. About a quarter of the
original matter's mass should end up as helium, only
about 0.01 per cent should end up as deuterium, and
even less should have become lithium. The rest
remained hydrogen. These ratios are indeed roughly
what we see today, providing strong support for the
Big Bang model.

> **I have this one
> little saying, when
> things get too heavy
> just call me helium,
> the lightest known
> gas to man.**
>
> Jimi Hendrix

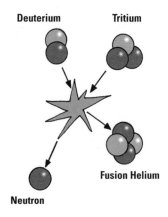

Deuterium **Tritium**

Fusion Helium

Neutron

Elementary puzzles

The theory of nucleosynthesis, worked out by physicists Ralph Alpher, Hans Bethe and George Gamow in the 1940s, did more than underpin the Big Bang. It fixed problems that had arisen by comparing predictions with the measured abundances of light elements in the stars. For years it had been known that helium and deuterium especially were more common than could be explained by stellar models at the time. The heavy elements are built up gradually in stars by nuclear fusion. Hydrogen burns to make helium, and chains of other reactions build up carbon, nitrogen and oxygen and a range of other elements. But helium is created only slowly, taking much of the lifetime of a star to make an appreciable amount. Deuterium is impossible to make in stars through normal fusion processes – it is only destroyed in stellar atmospheres. But by adding in the extra amounts that were created in the Big Bang, the maths could be fixed.

To measure the primordial ratios of light elements, astronomers seek out pristine regions of the universe. They look for slow-burning old stars that are relatively unpolluted by later heavy element production and recycling. Alternatively they look for ancient gas clouds that have changed little since the universe's early days. Lying in remote regions of intergalactic space, far from galactic pollutants, such clouds can be found when they absorb the light from distant objects such as bright quasars. The gas clouds' spectral fingerprints can give up their chemistry.

Matter measure The amount of deuterium created in the Big Bang is a particularly valuable measurement. Because it is made only through unusual nuclear reactions, its abundance depends sensitively on the original number of protons and neutrons in the young universe. The fact that deuterium is so rare implies that the density of these first nucleons was low, too low to be able to say that everything in the universe came from them. Other forms of exotic matter must also be present.

> **Things are the way they are because they were the way they were.**
> **Fred Hoyle**

HANS BETHE (1906–2005)

Born in Strasbourg in Alsace-Lorraine, Hans Bethe studied and taught theoretical physics at universities in Frankfurt, Munich and Tuebingen. When the Nazi regime came to power in 1933 he lost his university position and emigrated first to England, and then in 1935 to Cornell University in the USA. During the Second World War, he was head of the Theoretical Division at Los Alamos laboratory, where he performed calculations that were critical for developing the first atomic bombs. A prolific scientist, Bethe worked on many physics problems. He received the Nobel Prize for his theory of stellar nucleosynthesis, and also tackled other areas of nuclear and particle astrophysics. He later campaigned against nuclear weapons testing alongside Albert Einstein; he influenced the White House to sign the ban on atmospheric nuclear tests in 1963 and the 1972 Anti-Ballistic Missile Treaty, SALT I. Freeman Dyson called Bethe the 'supreme problem solver of the 20th century'.

Modern observations of galaxies, galaxy clusters and the cosmic microwave background hint that there are types of matter out there that are not based on protons and neutrons. This exotic matter is 'dark' and doesn't glow, and makes up the majority of the universe's mass. It might be made of unusual particles, such as neutrinos, or even black holes. The light element abundances show that normal matter makes up just a few per cent of the total mass of the universe.

the condensed idea
The first light elements

17 Antimatter

Fictional spaceships are often powered by 'antimatter drives', yet antimatter itself is real and has been made artificially on Earth. A mirror-image form of matter that has negative energy, antimatter cannot coexist with matter for long – both annihilate in a flash of energy if they come into contact. The fact that the universe is full of matter implies that antimatter is rare and hints at imbalances during the Big Bang.

Walking down the street you meet a replica of yourself. It is your antimatter twin. Do you shake hands? Antimatter was predicted in the 1920s and discovered in experiments in the 1930s. It is a mirror-image form of matter, where particles' charges, energies and other physical properties are all reversed in sign. So an anti-electron, called a positron, has the same mass as the electron but instead has a positive charge. Similarly, protons and other particles have opposite antimatter siblings.

Negative energy Creating an equation for the electron in 1928, British physicist Paul Dirac saw that it offered the possibility that electrons could have negative as well as positive energy. Just as the equation $x^2 = 4$ has the solutions $x = 2$ and $x = -2$, Dirac had two ways of solving his problem: positive energy was expected, associated with a normal electron, but negative energy made no sense. But rather than ignore this confusing term, Dirac suggested that such particles might actually exist. This complementary state of matter is 'anti' matter.

Antiparticles The hunt for antimatter began quickly. In 1932, Carl Anderson confirmed the existence of positrons experimentally. He was following the tracks of showers of particles produced by cosmic rays (energetic particles that crash into the atmosphere from space) when he

timeline

1928	1932
Dirac derives the existence of antimatter	Anderson detects the positron

saw the track of a positively charged particle with the electron's mass, the positron. So antimatter was no longer just an abstract idea but was real.

It took another two decades before the next antiparticle, the antiproton, was detected. Physicists built new particle-accelerating machines that used magnetic fields to increase the speeds of particles travelling through them. Such powerful beams of speeding protons produced enough energy to reveal the antiproton in 1955. Soon afterwards the antineutron was also found.

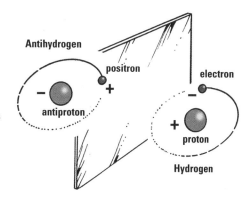

On Earth, physicists can create antimatter in particle accelerators, such as those at CERN in Switzerland or Fermilab near Chicago. When the beams of particles and antiparticles meet, they annihilate each other in a flash of pure energy. Mass is converted to energy according to Einstein's $E = mc^2$ equation. So if you met your antimatter twin it might not be such a good idea to throw your arms around them.

Universal asymmetries If antimatter were spread across the universe, these annihilation episodes would be occurring all the time. Matter and antimatter would gradually destroy each other in little explosions, mopping each other up. Because we don't see this, there cannot be much antimatter around. In fact normal matter is the only widespread form of particle we see, by a very large margin. So at the outset of the creation of the universe there must have been an imbalance such that more normal matter was created than its antimatter opposite.

In science one tries to tell people, in such a way as to be understood by everyone, something that no one ever knew before. But in poetry, it's the exact opposite.

Paul Dirac

1955	**1965**	**1995**
Antiprotons are detected	The first anti-nucleus is produced	Anti-hydrogen atoms are created

PAUL DIRAC 1902–84

Paul Dirac was a talented but shy British physicist. People joke that his vocabulary consisted of 'Yes', 'No', and 'I don't know'. He once said: 'I was taught at school never to start a sentence without knowing the end of it.' What he lacked in verbosity he made up for in his mathematical ability. His PhD thesis is famous for being impressively short and powerful, presenting a new mathematical description of quantum mechanics. He partly unified the theories of quantum mechanics and relativity theory, but he also is remembered for his outstanding work on the magnetic monopole and in predicting antimatter. When he was awarded the 1933 Nobel Prize, his first thought was to turn it down to avoid the publicity. But he gave in when told he would get even more publicity if he refused to accept it. Dirac did not invite his father to the ceremony, possibly because of strained relations after the suicide of his brother.

Like all mirror images, particles and their antiparticles are related through different kinds of symmetry. One is time. Because of their negative energy, antiparticles are equivalent mathematically to normal particles moving backwards in time. So a positron can be thought of as an electron travelling from future to past. The next symmetry involves charges and other quantum properties, which are reversed. A third symmetry regards motion through space. Motions are generally unaffected if we change the direction of coordinates marking out the grid of space. A particle moving left to right looks the same as one moving right to left, or is unchanged whether spinning clockwise or anticlockwise. This 'parity' symmetry is true of most particles, but there are a few for which it does not always hold. Neutrinos exist in only one form, as a left-handed neutrino, spinning in one direction; there is no such thing as a right-handed neutrino. The converse is true for antineutrinos, which are all right-handed. So parity symmetry can sometimes be broken, although a combination of charge conjugation and parity is conserved, called charge–parity symmetry for short.

> **The opposite of a correct statement is a false statement. But the opposite of a profound truth may well be another profound truth.**
>
> **Niels Bohr**

Just as chemists find that some molecules prefer to exist in one version, as a left-handed or right-handed structure, it is a major puzzle why the

> **For every one billion particles of antimatter there were one billion and one particles of matter. And when the mutual annihilation was complete, one billionth remained – and that's our present universe.**
>
> **Albert Einstein**

universe contains mostly matter and not antimatter. A tiny fraction – less than 0.01 per cent – of the stuff in the universe is made of antimatter. But the universe also contains forms of energy, including a great many photons. So it is possible that a vast amount of both matter and antimatter was created in the Big Bang, but then most of it was annihilated shortly afterwards. Only the tip of the iceberg now remains. A minuscule imbalance in favour of matter would be enough to explain its dominance now. Only 1 in every 10,000,000,000 (10^{10}) matter particles needed to survive a split second after the Big Bang, the remainder being annihilated. The leftover matter was likely preserved via a slight asymmetry in charge and parity.

The particles that may have been involved in this asymmetry, called X bosons, have yet to be found. These massive particles decay in an imbalanced way to give a slight overproduction of matter. X bosons may also interact with protons and cause them to decay, which would be bad news as it means that all matter will eventually disappear into a mist of even finer particles. But the good news is that the timescale for this happening is very long. That we are here and no one has ever seen a proton decay means that protons are very stable and must live for at least 10^{17}–10^{35} years, or billions of billions of billions of years, hugely longer than the lifetime of the universe so far. But this does raise the possibility that if the universe grows really old, then even normal matter might one day disappear.

the condensed idea
Mirror-image matter

18 Dark matter

Ninety per cent of the matter in the universe does not glow but is dark. Dark matter is detectable by its gravitational effect but hardly interacts with light waves or matter. Scientists think it may be in the form of MACHOs, failed stars and gaseous planets, or WIMPs, exotic subatomic particles. The hunt for dark matter is the wild frontier of physics.

Dark matter sounds exotic, and it may be, but its definition is quite mundane. Most of the things we see in the universe glow because they emit or reflect light. Stars twinkle by pumping out photons, and the planets shine by reflecting light from the Sun. Without that light, we simply would not see them. When the Moon passes into the Earth's shadow it is dark; when stars burn out they leave husks too faint to see; even a planet as big as Jupiter would be invisible if it was set free to wander far from the Sun. So it is, at first sight, perhaps not a big surprise that much of the stuff in the universe does not glow. It is dark matter.

Dark side Although we cannot see dark matter directly, we can detect its mass through its gravitational pull on other astronomical objects and also light rays. If we did not know the Moon was there, we could still infer its presence because its gravity would tug and shift the orbit of the Earth slightly. We have even used the gravity-induced wobble applied to a parent star to discover planets around distant stars.

In the 1930s, Swiss astronomer Fritz Zwicky realized that a nearby giant cluster of galaxies was behaving in a way that implied that its mass was much greater than the weight of all the stars in the galaxies within it. He inferred that some unknown dark matter accounted for 400 times as much material as luminous matter, glowing stars and hot gas, across the entire

timeline

1933	1975
Zwicky measures dark matter in the Coma cluster	Vera Rubin shows that galaxy rotation is affected by dark matter

cluster. The sheer amount of dark matter was a big surprise, implying that most of the universe was not in the form of stars and gas but something else. So what is this dark stuff? And where does it hide?

Mass is also missing from individual spiral galaxies. Gas in the outer regions rotates faster than it should do if the galaxy was only as heavy as the combined mass of stars within it. So such galaxies are more massive than expected by looking at the light alone. Again, the extra dark matter needs to be hundreds of times more abundant than the visible stars and gas. Dark matter is not only spread throughout galaxies, but its mass is so great that it dominates the motions of every star within them. Dark matter even extends beyond the stars, filling a spherical 'halo' or bubble around every flattened spiral galaxy disc.

Weight gain Astronomers have now mapped dark matter not only in individual galaxies but also in clusters containing thousands of galaxies bound together by mutual gravity, and in superclusters, chains of clusters of galaxies in a vast web that stretches across all of space. Dark matter features wherever there is gravity at work, on every scale. If we add up all the dark matter, we find that there is a thousand times more dark stuff than there is luminous matter.

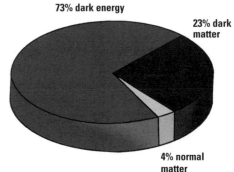

73% dark energy

23% dark matter

4% normal matter

The fate of the entire universe depends on its overall weight. Gravity's attraction counterbalances the expansion of the universe following the Big Bang explosion. There are three possible outcomes. Either the universe is so heavy that gravity wins and it eventually collapses back in on itself (a closed universe ending in a big crunch), or there is too little mass and it expands for ever (an open universe), or it is precisely balanced and the expansion gradually slows through gravity, but over such a long time that it never ceases. The latter seems the best case for our universe; it has precisely the right amount of matter to slow but never halt the expansion.

1998
Neutrinos inferred to
have a small mass

2000
MACHOs detected
in the Milky Way

Energy budget

Today we know that only about 4 per cent of the universe's matter is made up of baryons (normal matter comprising protons and neutrons). Another 23 per cent is exotic dark matter. We do know that this isn't made up of baryons. It is harder to say what it *is* made from, but it could be particles such as WIMPs. The rest of the universe's energy budget consists of something else entirely, dark energy (see p.82).

WIMPs and MACHOs What might dark matter be made of? First, it could be dark gas clouds, dim stars or unlit planets. These are called MACHOs, or Massive Compact Halo Objects. Alternatively the dark matter could be new kinds of subatomic particles, called WIMPs, short for Weakly Interacting Massive Particles, which would have virtually no effect on other matter or light.

Astronomers have found MACHOs roaming within our own galaxy. Because MACHOs are large, akin to the planet Jupiter, they can be spotted individually by their gravitational effect. If a large gas planet or failed star passes in front of a background star, its gravity bends the starlight around it. The bending focuses the light when the MACHO is right in front of the star, so the star appears much brighter for a moment as it passes. This is called 'gravitational lensing'.

In terms of relativity theory, the MACHO planet distorts space–time, like a heavy ball depressing a rubber sheet, which curves the light's wavefront around it (see p.93). Astronomers have looked for this brightening of stars by the passage of a foreground MACHO against millions of stars in the background, but have found only a few such flare-ups, too few to explain all the missing mass of the Milky Way.

❝The universe is made mostly of dark matter and dark energy, and we don't know what either of them is.❞
Saul Perlmutter

MACHOs are made of normal matter, or baryons, built of protons, neutrons and electrons. The tightest limit on the amount of baryons in the universe is given by tracking the heavy hydrogen isotope deuterium. Deuterium was only produced in the Big Bang itself and is not formed by stars afterwards, although it can be burned within them. So, by measuring the amount of deuterium in pristine gas clouds in space, astronomers can estimate the total number of protons and neutrons that were made in the Big Bang, because the mechanism for making deuterium is precisely known. This turns out to be just a few per cent of the mass of the entire universe. So the rest of the universe must be in some entirely different form, such as WIMPs.

The search for WIMPS is now the focus of attention. Because they are weakly interacting, these particles are intrinsically difficult to detect. One candidate is the neutrino. In the last decade physicists have measured its mass and found it to be very small but not zero. Neutrinos make up some of the universe's mass, but again not all. So there is still room for other more exotic particles out there waiting to be detected, some new to physics such as axions and photinos. Understanding dark matter may yet light up the world of physics.

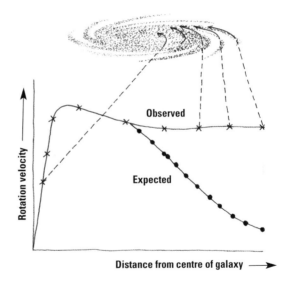

The outer parts of a spiral galaxy spin faster due to dark matter

the condensed idea
Dark side of the universe

19 Cosmic inflation

Why does the universe look the same in all directions? And why, when parallel light rays traverse space, do they remain parallel so that we see separate stars? We think that the answer is inflation – the idea that the baby universe swelled up so fast in a split second that its wrinkles smoothed out and its subsequent expansion balanced gravity exactly.

The universe we live in is special. When we look out into it we see clear arrays of stars and distant galaxies without distortion. It could so easily be otherwise. Einstein's general relativity theory describes gravity as a warped sheet of space and time upon which light rays wend their way along curved paths (see p.93). So, potentially, light rays could become scrambled, and the universe we look out on to could appear distorted like reflections in a hall of mirrors. But overall, apart from the odd deviation as they skirt a galaxy, light rays tend to travel more or less in straight lines right across the universe. Our perspective remains clear all the way to the visible edge.

Flatness Although relativity theory thinks of space–time as being a curved surface, astronomers sometimes describe the universe as 'flat', meaning that parallel light rays remain parallel no matter how far they travel through space, just as they would do if travelling along a flat plane. Space–time can be pictured as a rubber sheet; heavy objects that weigh down the sheet and rest in its dips represent gravity. In reality, space–time has more dimensions (at least four: three of space and one of time), but it is hard to imagine those. The fabric is also continually expanding, following the Big Bang explosion. The universe's geometry is such that the sheet remains mostly flat, like a tabletop, give or take some small dips and lumps here or there due to the patterns of matter. So light's path across the universe is relatively unaffected, bar the odd detour around a massive body.

 timeline

If there was too much matter, then everything would weigh the sheet down and it would eventually fold in on itself, reversing the expansion. In this scenario, parallel light rays would eventually converge. If there was too little matter weighing it down, then the space–time sheet would stretch and pull itself apart, and the parallel light rays would diverge as they crossed it. However, our real universe seems to be somewhere in the middle, with just enough matter to hold the universe's fabric together while expanding steadily. So the universe appears to be precisely poised.

Sameness Another feature of the universe is that it appears roughly the same wherever we look. The galaxies do not concentrate in one spot; they are littered in all directions. This might not seem that surprising at first, but it is unexpected. The puzzle is that the universe is so big that its opposite edges should not be able to communicate even at the speed of light. Though it has only existed for 14 billion years, it is more than 14 billion light years across in size, so light, even though it is travelling at the fastest speed attainable by any transmitted signal, has not had time to get from one side of the universe to the other. How then does one side of the universe know what the other side should look like? This

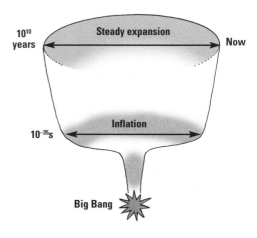

is the 'horizon problem', where the 'horizon' is the furthest distance that light has travelled since the birth of the universe, marking an illuminated sphere. There are regions of space that we cannot and will never see, because light from there has not yet had time to travel to us.

> **❛It is said that there's no such thing as a free lunch. But the universe is the ultimate free lunch.❜**
>
> **Alan Guth**

Geometry of the universe

From the latest observations of the microwave background, such as those of the Wilkinson Microwave Anisotropy Probe (WMAP) satellite in 2003 and 2006, physicists have been able to measure the shape of space–time right across the universe. By comparing the sizes of hot and cold patches in the microwave sky with the lengths predicted for them by Big Bang theory, they show that the universe is 'flat'. Even over a journey across the entire universe lasting billions of years, light beams that set out parallel will remain parallel.

Smoothness The universe is also quite smooth. Galaxies are spread fairly uniformly across the sky. If you squint, they form an even glow rather than clumping in a few big patches. Again this need not have been the case. Galaxies have grown over time due to gravity. They started out as just a slightly overdense spot in the gas left over from the Big Bang. That spot started to collapse due to gravity, forming stars and eventually building up a galaxy. The original overdense seeds of galaxies were set up by quantum effects, minuscule shifts in the energies of particles in the hot embryonic universe. But they could well have amplified to make large galaxy patches, like a cow's hide, unlike the widely scattered sea that we see. There are many molehills in the galaxy distribution, rather than a few giant mountain ranges.

Growth spurt The flatness, horizon and smoothness problems of the universe can all be fixed with one idea: inflation. Inflation was developed as a solution in 1981 by American physicist Alan Guth. The horizon problem, that the universe looks the same in all directions even though it is too large to know this, implies that it must at one time have been so small that light could communicate between all its regions. Because it is no longer like this, it must have then inflated quickly to the proportionately bigger universe we see now. But this period of inflation must have been extraordinarily rapid, much faster than the speed of light. The rapid expansion, doubling in size

> **❝It is rather fantastic to realize that the laws of physics can describe how everything was created in a random quantum fluctuation out of nothing.❞**
> **Alan Guth**

and doubling again and again in a split second, smeared out the slight density variations imprinted by quantum fluctuations, just as a printed pattern on an inflated balloon becomes fainter, and the universe became smooth. The inflationary process also fixed up the subsequent balance between gravity and the final expansion, proceeding at a much more leisurely pace thereafter. Inflation happened almost immediately after the Big Bang fireball (10^{-35} seconds afterwards).

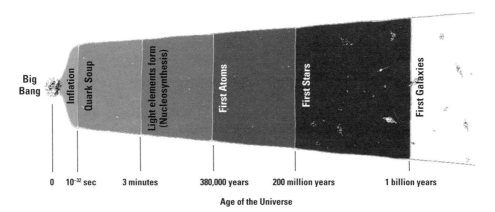

| Big Bang | Inflation | Quark Soup | Light elements form (Nucleosynthesis) | First Atoms | First Stars | First Galaxies |

0   10^{-32} sec   3 minutes   380,000 years   200 million years   1 billion years

Age of the Universe

Inflation has not yet been proven and its ultimate cause is not well understood – there are as many models as theorists – but understanding it is a goal of the next generation of cosmology experiments, which will involve the production of more detailed maps of the cosmic microwave background radiation and its polarization.

the condensed idea
Cosmic growth spurt

20 Cosmological constant

Albert Einstein believed that adding his cosmological constant into the equations of general relativity was his biggest blunder. The term allowed for the speeding up or slowing down of the rate of expansion of the universe to compensate gravity. Einstein did not need this number and abandoned it. However, new evidence in the 1990s required that it be reintroduced. Astronomers found that mysterious dark energy is causing the expansion of the universe to speed up, leading to the rewriting of modern cosmology.

Einstein thought that we lived in a steady state universe rather than one with a Big Bang. Trying to write down the equations for this, however, he ran into a problem. If you just had gravity, then everything in the universe would ultimately collapse into a point, perhaps a black hole. Obviously the real universe wasn't like that, so Einstein added another term to his theory to counterbalance gravity, a sort of repulsive 'anti-gravity' term. He introduced this purely to make the equations look right, not because he knew of such a force. But this formulation was immediately problematic.

If there was a counterforce to gravity, then just as untrammelled gravity could cause collapse, an anti-gravity force could as easily amplify to tear apart regions of the universe that were not held together by gravity's glue. Rather than allow such shredding of the universe, Einstein preferred to ignore his second repulsive term and admit he had made a mistake in introducing it. Other physicists also preferred to exclude it, relegating it to

timeline
1915
Einstein publishes the general
theory of relativity

> **It is to be emphasized, however, that a positive curvature of space is given by our results, even if the supplementary term [cosmological constant] is not introduced. That term is necessary only for the purpose of making possible a quasi-static distribution of matter.**
>
> **Albert Einstein**

history. Or so they thought. The term was not forgotten – it was preserved in the relativity equations, but its value, the cosmological constant, was set to zero to dismiss it.

Accelerating universe In the 1990s, two groups of astronomers were mapping supernovae in distant galaxies to measure the geometry of space when they found that distant supernovae appeared fainter than they should be. Supernovae, the brilliant explosions of dying stars, come in many types. Type Ia supernovae have a predictable brightness and so are useful for inferring distances. Just like the Cepheid variable stars that were used to measure the distances to galaxies to establish Hubble's law, the intrinsic brightness of Type Ia supernovae can be worked out from their light spectra so that it is possible to say how far away they must be. This all worked fine for supernovae that were nearby, but the more distant supernovae were too faint. It was as if they were further away from us than they should be.

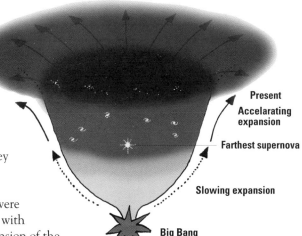

Present
Accelerating expansion

Farthest supernova

Slowing expansion

Big Bang

As more and more distant supernovae were discovered, the pattern of the dimming with distance began to suggest that the expansion of the

1929

Hubble shows space is expanding and Einstein abandons his constant

1998

Supernova data indicate the need for the cosmological constant

❝For 70 years, we've been trying to measure the rate at which the universe slows down. We finally do it, and we find out it's speeding up.❞

Michael S. Turner

universe was not steady, as in Hubble's law, but was accelerating. This was a profound shock to the cosmology community, and one that is still being disentangled today.

The supernova results fitted well with Einstein's equations, but only once a negative term was included by raising the cosmological constant from zero to about 0.7. The supernova results, taken with other cosmological data, such as the cosmic microwave background radiation pattern, showed that a new repulsive force counteracting gravity was needed. But it was quite a weak force. It is still a puzzle today why it is so weak, as there is no particular reason why it did not adopt a much larger value and perhaps completely dominate space over gravity. Instead it is very close in strength to gravity so has a subtle effect on space–time as we see it now. This negative energy term has been named 'dark energy'.

Dark energy The origin of dark energy is still elusive. All we know is that it is a form of energy associated with the vacuum of free space, creating a negative pressure in regions devoid of gravity-attracting matter and thus causing regions of empty space to inflate. We know its strength roughly from the supernova observations, but we do not know much more. We don't know if it truly is a constant – whether it always takes the same value right across the universe and for all time (as do gravity and the speed of

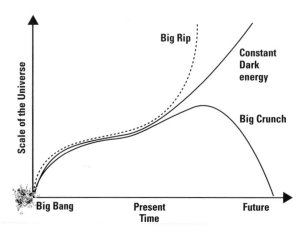

light) – or whether its value changes with time so that it may have had a different value just after the Big Bang compared with now or in the future. In its more general form it has also been called 'quintessence', or the fifth force, encompassing all the possible ways its strength could change with time. But it is still not known how this elusive force manifests itself or how it arises within the physics of the Big Bang. It is a hot topic of study for physicists.

> **❝It [dark energy] seems to be something that is connected to space itself and unlike dark matter which gravitates this has an effect which is sort of the opposite, counter to gravity, it causes the universe to be repulsed by itself.❞**
>
> **Brian Schmidt**

Nowadays we have a much better understanding of the geometry of the universe and what it is made up of. The discovery of dark energy has balanced the books of cosmology, making up the difference in the energy budget of the whole universe. So we now know that it is 4 per cent normal baryonic matter, 23 per cent exotic non-baryonic matter, and 73 per cent dark energy. These numbers add up to about the right amount of stuff for the balanced 'Goldilocks universe', close to the critical mass where it is neither open nor closed.

The mysterious qualities of dark energy, however, mean that even knowing the total mass of the universe, its future behaviour is hard to predict because it depends on whether or not the influence of dark energy increases in the future. If the universe is accelerating, then at this point in time, dark energy is only just as significant as gravity in dominating it. But at some point the acceleration will pick up and the faster expansion will overtake gravity. So the universe's fate may well be to expand for ever, faster and faster. Some scary scenarios have been proposed – once gravity is outstripped, then tenuously held-together massive structures will disconnect and fly apart; eventually even galaxies themselves will break up, then stars will be evaporated into a mist of atoms. Ultimately the negative pressure could strip atoms, leaving only a grim sea of subatomic particles.

Nevertheless, although cosmology's jigsaw is fitting together now, and we have measured a lot of the numbers that describe the geometry of the universe, there are still some big unanswered questions. We just don't know what 95 per cent of the stuff in the universe consists of, nor what this new force of quintessence really is. So it is not yet time to sit back and rest on our laurels. The universe has kept its mystery.

the condensed idea
The fifth force

21 Mach's principle

Due to gravity, everything in the universe attracts and is attracted by everything else. Ernst Mach, Austrian philosopher and physicist, thought about why it is that objects far away affect how things move, and spin, nearby – how the distant stars even tug upon a child on a merry-go-round. His principle, that 'mass there influences inertia here', grew from asking how you can tell if something is moving or not.

If you have ever sat in a train at a station and seen through the window a neighbouring carriage pull away from yours, you will know that sometimes it is hard to tell whether it is your own train leaving the station or the other arriving. The same thing led us to think incorrectly for centuries that the Sun orbited the Earth. Is there a way that you could measure for sure which one is in motion?

Mach grappled with this question in the 19th century. He was treading in the footsteps of Isaac Newton, who had believed, unlike Mach, that space was an absolute backdrop. Like graph paper, Newton's space contained an engraved set of coordinates, and he mapped all motions as movements with respect to that grid. Mach, however, disagreed, arguing instead that motion was only meaningful if measured with respect to another object, rather than the grid. What does it mean to be moving if not relative to something else?

In this sense Mach, who was influenced by the earlier ideas of Newton's competitor Gottfried Leibniz, was a forerunner to Albert Einstein in preferring to think that only relative motions made sense. Mach argued that because a ball rolls in the same way whether it is in France or Australia, the grid of space is irrelevant. The only thing that can conceivably affect how the ball rolls is gravity. On the Moon it might well

timeline

*c.*335 BC	1640
Aristotle states that objects move due to the action of forces	Galileo formulates the principle of inertia

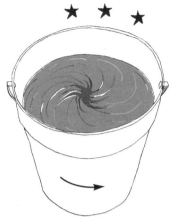

roll differently because the gravitational force pulling on its mass is weaker there. Because every object in the universe exerts a gravitational pull on every other, each object will feel each other's presence through their mutual attractions. So motion must ultimately depend on the distribution of matter, or its mass, not on the properties of space itself.

Mass What exactly is mass? It is a measure of how much matter an object contains. The mass of a lump of metal would be equal to the sum of the masses of all the atoms in it. Mass is subtly different from weight. Weight is a measure of the force of gravity pulling a mass down – an astronaut weighs less on the Moon than on Earth because the gravitational force exerted by the smaller Moon is less. But the astronaut's mass is the same – the number of atoms he contains has not changed. According to Albert Einstein, who showed that energy and mass are interchangeable, mass can be converted into pure energy. So mass is, ultimately, energy.

Inertia Inertia, named after the Latin word for 'laziness', is very similar to mass but tells us how hard it is to move something by applying a force. An object with large inertia resists movement. Even in outer space a massive object takes a large force to move it. A giant rocky asteroid on collision course with the Earth may need a huge shove to deflect it, whether it is created by a nuclear explosion or by a smaller force applied for a longer time. A smaller spacecraft, with less inertia than the asteroid, might be manoeuvred easily with tiny jet engines.

It was the 17th-century Italian astronomer Galileo Galilei who proposed the principle of inertia: if an object is left alone, and no forces are applied to it, then its state of motion is unchanged. If it is moving, it continues to move at the same speed and in the same direction. If it is standing still it continues to do so. Newton refined this idea to form his first law of motion.

Newton's bucket Newton also codified gravity. He saw that masses

1687	**1893**	**1905**
Newton publishes his bucket argument	Mach publishes 'the science of mechanics'	Einstein publishes the special theory of relativity

attract one another. An apple falls from a tree to the ground because it is attracted by the Earth's mass. Equally, the Earth is attracted by the apple's mass, but we would be hard pressed to measure the microscopic shift of the whole Earth towards the apple.

Newton proved that the strength of gravity falls off quickly with distance, so the Earth's gravitational force is much weaker if we are floating high above it rather than on its surface. But nevertheless we would still feel the reduced pull of the Earth. The further away we go, the weaker it would get, but it could still tweak our motion. In fact, all objects in the universe may exert a tiny gravitational pull that might subtly affect our movement.

Newton tried to understand the relationships between objects and movement by thinking about a spinning bucket of water. At first when the bucket is turned, the water stays still, even though the bucket moves. Then the water starts to spin as well. Its surface dips as the liquid tries to escape by creeping up the sides but it is kept in place by the bucket's confining force. Newton argued that the water's rotation could only be understood if seen in the fixed reference frame of absolute space, against its grid. We could tell if the bucket was spinning just by looking at it because we would see the forces at play on it producing the concave surface of the water.

Centuries later Mach revisited the argument. What if the water-filled bucket were the only thing in the universe? How could you know it was the bucket that was rotating? Couldn't you equally well say the water was rotating relative to the bucket? The only way to make sense of it would be to place another object into the bucket's universe, say the wall of a room, or

ERNST MACH 1838–1916

As well as for Mach's principle, Austrian physicist Ernst Mach is remembered for his work in optics and acoustics, the physiology of sensory perception, the philosophy of science and particularly his research on supersonic speed. He published an influential paper in 1877 that described how a projectile moving faster than the speed of sound produces a shock wave, similar to a wake. It is this shock wave in air that causes the sonic boom of supersonic aircraft. The ratio of the speed of the projectile, or jet plane, to the speed of sound is now called the Mach number, such that Mach 2 is twice the speed of sound.

> **❝Absolute space, of its own nature without reference to anything external, always remains homogenous and immovable.❞**
> **Isaac Newton, 1687**

even a distant star. Then the bucket would clearly be spinning relative to that. But without the frame of a stationary room, and the fixed stars, who could say whether it was the bucket or the water that rotates? We experience the same thing when we watch the Sun and stars arc across the sky. Is it the stars or the Earth that is rotating? How can we know?

According to Mach, and Leibniz, motion requires external reference objects for us to make sense of it, and therefore inertia as a concept is meaningless in a universe with just one object in it. So if the universe were devoid of any stars, we'd never know that the Earth was spinning. The stars tell us we're rotating relative to them.

The ideas of relative versus absolute motion expressed in Mach's principle have inspired many physicists since, notably Einstein, who coined the name 'Mach's principle'. Einstein used the idea that all motion is relative to build his theories of special and general relativity. He also solved one of the outstanding problems with Mach's ideas: rotation and acceleration must create extra forces, but where were they? Einstein showed that if everything in the universe were rotating relative to the Earth, we would indeed experience a small force that would cause the planet to wobble in a certain way.

The nature of space has puzzled scientists for millennia. Modern particle physicists think it is a seething cauldron of subatomic particles being continually created and destroyed. Mass, inertia, forces and motion may all in the end be manifestations of a bubbling quantum soup.

the condensed idea
Mass matters for motion

22 Special relativity

By thinking about relative motions, Albert Einstein showed in 1905 that strange effects happen when things move very quickly. Watching an object approach light speed, you'd see it become heavier, contract in length and age more slowly. That's because nothing can travel faster than the speed of light, so time and space distort to compensate when approaching this universal speed limit.

It is true that 'in space no one can hear you scream': sound waves ring though air, but their vibrations cannot be transmitted where there are no atoms, whereas light can spread through empty space, as we know because we see the Sun and stars. Is space filled with a special medium, a sort of electric air, through which electromagnetic waves propagate? Physicists at the end of the 19th century thought so, believing that space was effused with a gas or 'ether' through which light could radiate.

Light speed In 1887, however, a famous experiment proved the ether did not exist. Because the Earth moves around the Sun, its position in space is always changing. Albert Michelson and Edward Morley devised an ingenious experiment that would detect movement against it if the ether were fixed. They compared two beams of light travelling different paths, fired at right angles to one another and reflected back off identically faraway mirrors. Just as a swimmer takes less time to travel across a river from one bank to the other and back than to swim the same distance upstream against the current and downstream with it, they expected a similar result for light. The river current mimics the motion of the Earth through the ether. But there was no such difference – the light beams returned to their starting points at exactly the same time. No matter which direction the light

timeline

1887
Michelson and Morley are
unable to verify the ether

1893
Mach publishes 'the
science of mechanics'

> **The introduction of a light-ether will prove to be superfluous since . . . neither will a space in absolute rest endowed with special properties be introduced nor will a velocity vector be associated with a point of empty space in which electromagnetic processes take place.**
>
> **Albert Einstein**

travelled, and how the Earth was moving, the speed of light remained unchanged. Light's speed was unaffected by motion. The experiment proved the ether did not exist – but it took Einstein to realize this.

Just like Mach's principle (see p.87), this meant that there was no fixed background grid against which objects moved. Unlike water waves or sound waves, light appeared to always travel at the same speed. This was odd and quite different from our usual experience where velocities add together. If you are driving in a car at 50 km/h and another passes you at 65 km/h, it is as if you are stationary and the other vehicle is travelling at 15 km/h past you. But even if you were rushing at hundreds of km/h, light would still travel at the same speed. It is exactly 300 million metres per second whether you are shining a torch from your seat in a fast jet plane or from the saddle of a bicycle.

It was this fixed speed of light that puzzled Albert Einstein in 1905, leading him to devise his theory of special relativity. Then an unknown Swiss patent clerk, Einstein worked out the equations from scratch in his spare moments. Special relativity was the biggest breakthrough since Newton and revolutionized physics. Einstein started with the assumption that the speed of light is a constant value, and appears the same for any observer no matter how fast they are moving. If the speed of light does not change, reasoned Einstein, something else must change to compensate.

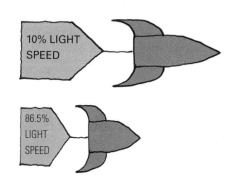

1905	1915	1971
Einstein publishes the special theory of relativity	Einstein publishes the theory of general relativity	Time dilation is demonstrated by flying clocks in planes

Twin paradox

Imagine if time dilation applied to humans. Well it could. If your identical twin was sent off into space on a rocket ship, fast enough and for long enough, they would age more slowly than you on Earth. On their return, they might find you to be elderly when they are still a sprightly youth. Although this seems impossible, it is not really a paradox, because the space-faring twin would experience powerful forces that permit such a change to happen. Because of this time shift, events that appear simultaneous in one frame may not appear so in another. Just as time slows, so lengths contract also. The object or person moving at that speed would not notice either effect; it would just appear so to another viewer.

Space and time Following ideas developed by Edward Lorenz, George Fitzgerald and Henri Poincaré, Einstein showed that space and time must distort to accommodate the different viewpoints of observers travelling close to the speed of light. The three dimensions of space and one of time made up a four-dimensional world in which Einstein was able to exercise his vivid imagination. Speed is distance divided by time, so to prevent anything from exceeding the speed of light, distances must shrink and time slow down to compensate. So a rocket travelling away from you at near light speed looks shorter and experiences time more slowly than you do.

Einstein worked out how the laws of motion could be rewritten for observers travelling at different speeds. He ruled out the existence of a stationary frame of reference, such as the ether, and stated that all motion was relative, with no privileged viewpoint. If you are sitting on a train and see the train next to you moving, you may not know whether it is your train or the other one pulling out. Moreover, even if you can see that your train is stationary at the platform you cannot assume that you are immobile, just that you are not moving relative to that platform. We do not feel the motion of the Earth around the Sun; similarly, we never notice the Sun's path across our own galaxy, or our Milky Way being pulled towards the huge Virgo cluster of galaxies beyond it. All that is experienced is relative motion, between you and the platform or the Earth spinning against the stars.

❝The most incomprehensible thing about the world is that it is at all comprehensible.❞

Albert Einstein

Einstein called these different viewpoints inertial frames. Inertial frames are spaces that move relative to one another at a constant speed, without experiencing accelerations or forces. So sitting in a car travelling at 50 km/h you are in one inertial frame, and you feel just the same as if you were in a train travelling at 100 km/h (another inertial frame) or a jet plane travelling at 500 km/h (yet another). Einstein stated that the laws of physics are the same in all inertial frames. If you dropped your pen in the car, train or plane, it would fall to the floor in the same way.

It is impossible to travel faster than the speed of light, and certainly not desirable, as one's hat keeps blowing off.

Woody Allen

Slower and heavier Turning next to relative motions near the speed of light, the maximum speed practically attainable by matter, Einstein predicted that time would slow down. Time dilation expressed the fact that clocks in different moving inertial frames may run at different speeds. This was proved in 1971 by sending four identical atomic clocks on scheduled flights twice around the world, two flying eastwards and two westwards. Comparing their times with a matched clock on the Earth's surface in the United States, the moving clocks had each lost a fraction of a second compared with the grounded clock, in agreement with Einstein's special relativity.

Another way that objects are prevented from passing the light-speed barrier is that their mass grows, according to $E = mc^2$. An object would become infinitely large at light speed itself, making any further acceleration impossible. And anything with mass cannot reach the speed of light exactly, but only approach it, as the closer it gets, the heavier and more difficult to accelerate it becomes. Light is made of mass-less photons so these are unaffected.

Einstein's special relativity was a radical departure from what had gone before. The equivalence of mass and energy was shocking, as were all the implications for time dilation and mass. Although Einstein was a scientific nobody when he published his ideas, they were read by top physicist Max Planck, and it is perhaps because of his adoption of the theory that it became accepted and not sidelined. Planck saw the beauty in Einstein's equations, catapulting him to global fame.

the condensed idea
Motion is relative

23 General relativity

Incorporating gravity into his theory of special relativity, Albert Einstein's theory of general relativity revolutionized our view of space and time. Going beyond Newton's laws, it opened up a universe of black holes, worm holes and gravitational lenses.

Imagine a person jumping off a tall building, or parachuting from a plane, being accelerated towards the ground by gravity. Einstein realized that in this state of free fall they did not experience gravity. In other words they were weightless. Trainee astronauts today re-create the zero-gravity conditions of space in just this way, by flying in a passenger jet (attractively nicknamed the Vomit Comet) in a path that mimics a roller coaster. When the plane flies upwards, the passengers are glued to their seats as they experience even stronger forces of gravity. But when it tips forwards and plummets downwards, they are released from gravity's pull and can float in the body of the aircraft.

Acceleration Einstein recognized that this acceleration was equivalent to the force of gravity. So, just as special relativity describes what happens in reference frames, or inertial frames, moving at some constant speed relative to one another, gravity was a consequence of being in a reference frame that is accelerating. He called this the happiest thought of his life.

Over the next few years Einstein explored the consequences. Talking through his ideas with trusted colleagues and using the latest mathematical formalisms to encapsulate them, he pieced together the full theory of gravity that he called general relativity. The year he published the work, 1915, proved especially busy, and almost immediately he revised it several times. His peers were astounded by his progress. The theory even produced

timeline

1687	1905
Newton proposes his theory of gravitation	Einstein publishes special theory of relativity

❝Time and space and gravitation have no separate existence from matter.❞
Albert Einstein

bizarre testable predictions, including the idea that light could be bent by a gravitational field, and also that Mercury's elliptical orbit would rotate slowly because of the gravity of the Sun.

Space–time In general relativity theory, the three dimensions of space and one of time are combined into a four-dimensional space–time grid, or metric. Light's speed is still fixed, and nothing can exceed it. When moving and accelerating, it is this space–time metric that distorts to maintain the fixed speed of light.

General relativity is best imagined by visualizing space–time as a rubber sheet stretched across a hollow tabletop. Objects with mass are like weighted balls placed on the sheet. They depress space–time around them. Imagine you place a ball representing the Earth on the sheet. It forms a depression in the rubber plane in which it sits. If you then threw in a

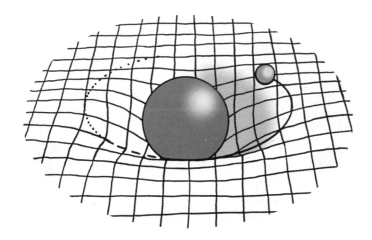

1915
Einstein publishes general
theory of relativity

1919
Eclipse observations verify
Einstein's theory

1960s
Evidence for black holes
seen in space

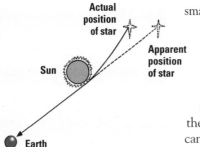

Actual position of star

Apparent position of star

Sun

Earth

smaller ball, say as an asteroid, it would roll down the slope towards the Earth. This shows how it feels gravity. If the smaller ball was moving fast enough and the Earth's dip was deep enough, then just as a daredevil cyclist can ride around an inclined track, that body would maintain a moon-like circular orbit. You can think of the whole universe as a giant rubber sheet. Every one of the planets and stars and galaxies causes a depression that can attract or deflect passing smaller objects, like balls rolling over the contours of a golf course.

Einstein understood that because of this warping of space–time, light would be deflected if it passed near a massive body, such as the Sun. He predicted that the position of a star observed just behind the Sun would shift a little because light from it is bent as it passes the Sun's mass. On 29 May 1919, the world's astronomers gathered to test Einstein's predictions by observing a total eclipse of the Sun. It proved one of his greatest moments, showing that the theory some thought crazy was in fact close to the truth.

Warps and holes The bending of light rays has now been confirmed with light that has travelled right across the universe. Light from very distant galaxies clearly flexes when it passes a very massive region such as a giant cluster of galaxies or a really big galaxy. The background dot of light is smeared out into an arc. Because this mimics a lens, the effect is known as gravitational lensing. If the background galaxy is sitting right behind the heavy intervening object, then its light is smeared out into a complete circle, called an Einstein ring. Many beautiful photographs of this spectacle have been taken with the Hubble Space Telescope.

Einstein's theory of general relativity is now widely applied to modelling the whole universe. Space–time can be thought of like a landscape, complete with hills, valleys and potholes. General relativity has lived up to all

> **❝A man sits with a pretty girl for an hour, it seems like a minute. He sits on a hot stove for a minute, it's longer than any hour. That is relativity.❞**
> **Albert Einstein**

gravity waves

Another aspect of general relativity is that waves can be set up in the space–time sheet, radiating especially from black holes and dense spinning compact stars like pulsars. Astronomers have seen pulsars' spin decreasing, so they expect that this energy will have been lost to gravity waves, but the waves have not yet been detected. Physicists are building giant detectors on Earth and in space that use the expected rocking of extremely long laser beams to spot the waves as they pass by. If gravity waves were detected, then this would be another coup for Einstein's general relativity theory.

observational tests so far. The regions where it is tested most are ones where gravity is especially strong, or perhaps very weak.

Black holes (see p.96) are extremely deep wells in the space–time sheet. They are so deep and steep that anything that comes close enough can fall in, even light. They mark holes, or singularities, in space–time. Space–time may also warp into worm holes, or tubes, but no one has actually seen such a thing yet.

At the other end of the scale, where gravity is very weak, it might be expected to break up eventually into tiny quanta, similar to light that is made up of individual photon building blocks. But no one has yet seen any graininess in gravity. Quantum theories of gravity are being developed, but without evidence to back it up, the unification of quantum theory and gravity is elusive. This hope occupied Einstein for the rest of his career, but even he did not manage it and the challenge still stands.

the condensed idea
Warped space–time

24 Black holes

Falling into a black hole would not be pleasant, having your limbs torn asunder and all the while appearing to your friends to be frozen in time just as you fell in. Black holes were first imagined as frozen stars whose escape velocity exceeds that of light, but are now considered as holes or 'singularities' in Einstein's space–time sheet. Not just imaginary, giant black holes populate the centres of galaxies, including our own, and smaller ones punctuate space as the ghosts of dead stars.

If you throw a ball up in the air, it reaches a certain height and then falls back down. The faster you fling it, the higher it goes. If you hurled it fast enough it would escape the Earth's gravity and whiz off into space. The speed that you need to reach to do this, called the 'escape velocity', is 11 km/s (or about 25,000 mph). A rocket needs to attain this speed if it is to escape the Earth. The escape velocity is lower if you are standing on the smaller Moon: 2.4 km/s would do. But if you were standing on a more massive planet, then the escape velocity rises. If that planet was heavy enough, the escape velocity could reach or exceed the speed of light itself, and so not even light could escape its gravitational pull. Such an object, which is so massive and dense that not even light can escape it, is called a black hole.

Event horizon The black hole idea was developed in the 18th century by geologist John Michell and mathematician Pierre-Simon Laplace. Later, after Einstein had proposed his relativity theories, Karl Schwarzschild worked out what a black hole would look like. In Einstein's theory of general relativity, space and time are linked and behave together like a vast rubber sheet. Gravity distorts the sheet according to an object's mass. A heavy planet rests in a dip in space–time, and its gravitational pull is

timeline

1784
Michell deduces the possibility of dark stars

1930s
Existence of frozen stars predicted

equivalent to the force felt as you roll into the dip, perhaps warping your path or even pulling you into orbit.

So what then is a black hole? It would be a pit that is so deep and steep that anything that comes close enough to it falls straight in and cannot return. It is a hole in the sheet of space–time, like a basketball net (from which you will never get your ball back).

If you pass far from a black hole, your path might curve towards it, but you needn't fall in. But if you pass too close to it, then you will spiral in. The same fate would even befall a photon of light. The critical distance that borders these two outcomes is called the 'event horizon'. Anything that falls within the event horizon, including light, plummets into the black hole.

Falling into a black hole has been described as being 'spaghetti-fied'. Because the sides are so steep, there is a very strong gravity gradient within it. If you were to fall into one feet first, and let's hope you never do, your feet would be pulled with more force than your head and so your body would be stretched as if it was on a rack. Add to that any spinning motion, and you would be pulled out like chewing gum into a scramble of spaghetti. Not a nice way to go. Some scientists have thought about ways of protecting anyone unlucky enough to accidentally stumble into a black hole. One way, apparently, is to don a leaden life-saver ring. If the ring was heavy and dense enough, it would counteract the gravity gradient and preserve your shape, and life.

Frozen stars The name 'black hole' was coined in 1967 by John Wheeler as a catchier alternative to describe a frozen star. Frozen stars were predicted in the 1930s by Einstein and Schwarzschild's theories. Because of the weird behaviour of space and time close to the event horizon, glowing matter falling in would seem to slow down as it does so, due to the light

1965	**1967**	**1970s**
Quasars discovered	Wheeler renames frozen stars as black holes	Hawking proposes that black holes evaporate

Evaporation

Strange as it may sound, black holes eventually evaporate. In the 1970s, Stephen Hawking suggested that they are not completely black but radiate particles due to quantum effects. Mass is gradually lost in this way and so the black hole shrinks until it disappears. The black hole's energy continually creates pairs of particles and their corresponding antiparticles. If this happens near the event horizon, then sometimes one of the particles might escape even if the other falls in. To an outside eye the black hole seems to emit particles, called Hawking radiation. This radiated energy then causes the hole to diminish. This idea is still based in theory, and no one really knows what happens to a black hole. The fact that they are relatively common suggests that this process takes a long time, so black holes hang around.

waves taking longer and longer to reach an observer looking on. As the material passes the event horizon, this outside observer sees time actually stop so that the matter appears to be frozen at the time it crosses the horizon. Hence, the star seems to freeze just at the point of collapsing into the event horizon, as predicted.

Astrophysicist Subrahmanyan Chandrasekhar predicted that stars more than 1.4 times the Sun's mass would ultimately collapse into a black hole; however, due to the laws of quantum physics, we now know that white dwarfs and neutron stars will prop themselves up, so stars of more than three times the Sun's mass are needed for black holes to form. Evidence of these frozen stars or black holes was not discovered until the 1960s.

If black holes suck in light, how can we see them to know they are there? There are two ways. First, you can spot them because of the way they pull other objects towards them. And second, as gas falls into them it can heat up and glow before it disappears. The first method has been used to identify

The black holes of nature are the most perfect macroscopic objects there are in the universe: the only elements in their construction are our concepts of space and time.

Subrahmanyan Chandrasekhar

> **God not only plays dice, but also sometimes throws them where they cannot be seen.**
> **Stephen Hawking**

a black hole lurking in the centre of our own galaxy. Stars that pass close to it have been seen to whip past it and be flung out on elongated orbits. The Milky Way's black hole has a mass of a million Suns, squashed into a region of radius just 10 million kilometres (30 light seconds) or so. Black holes that lie in galaxies are called supermassive black holes. We don't know how they formed, but they seem to affect how galaxies grow so might have been there from day one, or perhaps grew from millions of stars collapsing into one spot.

The second way to see a black hole is by the light coming from hot gas that is fired up as it falls in. Quasars, the most luminous things in the universe, shine due to gas being sucked into supermassive black holes in the centres of distant galaxies. Smaller black holes, just a few solar masses, can also be identified by X-rays shining from gas falling towards them.

Worm holes What lies at the bottom of a black hole in the space–time sheet? Supposedly they just end in a sharp point, or truly are holes, punctures in the sheet. But theorists have asked what might happen if they joined another hole. You can imagine that two nearby black holes might appear as long tubes dangling from the space–time sheet. If the tubes were joined together, then you could imagine a worm hole being formed between the mouths of the two black holes. Armed with your 'life-saver', you might be able to jump into one black hole and pop out of another. This idea has been used a lot in science fiction for transport across time and space. Perhaps the worm hole could flow through to an entirely different universe. The possibilities for rewiring the universe are endless, but don't forget your life-saver ring.

the condensed idea
Light traps

25 Particle astrophysics

Space is littered with particles, accelerated to immense energies by cosmic magnetic fields in the same way that Earth-bound physicists attempt with their modest man-made machines. The detection of cosmic rays, neutrinos and other exotic particles from space will help us explain what the universe is made of.

Since the ancient Greeks, we have thought that atoms were the basic building blocks of the universe. We now know better. Atoms can be dissected and are made up of lightweight negatively charged electrons, orbiting around a positively charged nucleus made of protons and neutrons. Even these particles can be split apart, and modern physics has revealed a zoo of fundamental particles, which formed in the Big Bang into the universe.

Unpeeling atoms Electrons were first liberated from atoms in the laboratory in 1887 by Joseph John Thomson, who fired an electric current through a gas-filled glass tube. Not long afterwards, in 1909, Ernest Rutherford discovered the nucleus – named after the Latin word for the kernel of a nut. Firing a stream of alpha particles (a form of radiation consisting of two protons and two neutrons) at a thin sheet of gold foil, he was surprised to find that a small fraction bounced straight back at him, having hit something compact and hard in the centre of the gold atom.

By isolating the nuclei of hydrogen, Rutherford identified protons in 1918. But matching up the charges and weights of other elements proved harder.

timeline

400 BC	1887	1909
Democritus proposes the idea of atoms	Thomson discovers the electron	Rutherford performs gold foil experiment

> **❝It was almost as incredible as if you fired a 15-inch shell at a piece of tissue paper and it came back to hit you.❞**
>
> ### Ernest Rutherford

In the early 1930s, James Chadwick found the missing ingredient – the neutron, a neutral particle with about the same mass as the proton. The various weights of elements, including those with odd weights called isotopes, could now be explained. A carbon-12 atom, for instance, contains six protons and six neutrons in the nucleus (to give it a mass of 12 atomic units) and six orbiting electrons, whereas carbon-14 is heavier because it has an extra two neutrons.

The nucleus is tiny. A hundred thousand times smaller than an atom, it is only a few femtometres (10^{-15} metres, or one ten million billionth of a metre) across. If the atom were scaled up to the diameter of the Earth, the nucleus at the centre would be just 10 kilometres wide, or the length of Manhattan.

Standard model As more was learned from radioactivity about how nuclei broke apart (via fission) or joined together (via fusion), other phenomena needed explanation. The burning of hydrogen into helium in the Sun, via fusion, implicated another particle, the neutrino, which transforms protons into neutrons. In 1930, the neutrino's existence was inferred to explain the decay of a neutron into a proton and electron – beta radioactive decay. The neutrino itself, having virtually no mass, was not discovered until 1956.

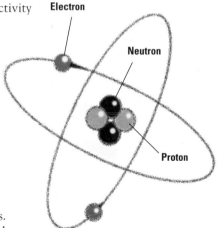

Electron

Neutron

Proton

In the 1960s, physicists realized that protons and neutrons were not the smallest building blocks: they hosted even smaller particles within them, called quarks. Quarks come in threes. They have three 'colours', red, blue

1918	**1932**	**1956**	**1960s**	**1995**
Rutherford isolates the proton	Chadwick discovers the neutron	Neutrino detected	Quarks are proposed	The top quark is found

and green, and also come in six 'flavours', as three pairs of increasing mass. The lightest are the 'up' and 'down' quarks; next come 'strange' and 'charm' quarks; finally, the 'top' and 'bottom' quarks are the heaviest pair. These unusual names were chosen by physicists to express the properties of the quarks, which are without precedent. Quarks cannot exist for long on their own, and must always be locked together in combinations that are colour neutral overall (exhibiting no colour charge). Possibilities include threesomes called baryons ('bary' means heavy), including normal protons and neutrons, or quark–antiquark pairs (called mesons). Three quarks are needed to make up a proton (two ups and a down) or a neutron (two downs and an up).

The next basic class of particles, the leptons, is related to and includes electrons. Again there are three generations with increasing masses: electrons, muons and taus. Muons are 200 times heavier than an electron and taus 3,700 times heavier. Leptons all have a single negative charge. They also have an associated particle called a neutrino (electron-, muon- and tau-neutrino) that has no charge. Neutrinos have almost no mass and do not interact much with anything. They can travel right through the Earth without being noticed, so are difficult to catch.

Fundamental forces are mediated by the exchange of particles. Just as the electromagnetic wave can also be thought of as a stream of photons, the weak nuclear force can be thought of as being carried by particles called W and Z bosons while the strong nuclear force is transmitted via gluons. Gravity isn't yet included in this standard model of particle physics described here, although physicists are trying.

Particle smashing Particle physics has been described as taking an intricate Swiss watch and smashing it up with a hammer, then looking at the shards to work out how it operates. Particle accelerators on Earth use giant magnets to accelerate particles to extremely high speeds and then smash those particle beams either into a target or into another oppositely directed beam. At modest speeds, the particles break apart a little and the lightest generations of particles are released. Because mass means energy, you need a higher-energy particle beam to release the heavier particles.

The particles produced are identified from photographs of their tracks. As they pass through a magnetic field, positive-charged particles swerve one way and negative ones the other. The mass of the particle also dictates how fast it shoots through the detector and how much its path is curved by the magnetic field. So light particles barely curve and heavier particles may even spiral into loops. By mapping their characteristics in the detector, and comparing them with what they expect from their theories, physicists can tell what each particle is.

Cosmic rays In space, particles are produced through similar processes to those in accelerators on Earth. Wherever there are strong magnetic fields – such as in the middle of our galaxy, in a supernova explosion or in jets blasted out from a black hole – particles can reach incredible energies, sometimes travelling at close to the speed of light. Anti-particles might also be created, raising the possibility that their annihilation could be observed when they come into contact with normal matter.

Cosmic rays are space-borne particles that crash into our atmosphere. As they collide with air molecules, they shatter and produce a cascade of further smaller particles, some of which reach the ground. These particle showers can be picked up as flashes on detectors on the Earth's surface. By measuring the characteristic energies of cosmic rays, and the directions from which they come, astronomers hope to understand their origin.

Neutrinos are also sought with anticipation because they are a candidate to make up the dark-matter budget of the universe. Because they hardly interact with anything, though, they are difficult to detect. To do so, physicists think big – they use the whole Earth as a detector. Neutrinos passing right through the Earth will occasionally be slowed down, and then vast arrays of detectors will be waiting, including new ones within the ice of Antarctica and in the Mediterranean sea. Other underground experiments placed deep in mines will ensnare different types of particles. Through such imaginative means, astronomers may learn in decades to come what makes up our universe.

> **Nothing exists except atoms and empty space; everything else is opinion.**
> **Democritus**

the condensed idea
Cosmic accelerator

26 The 'God particle'

While walking in the Scottish Highlands in 1964, physicist Peter Higgs thought of a way to give particles their mass. He called this his 'one big idea'. Particles seem more massive because they are slowed while swimming through a force field, now known as the Higgs field. The property of mass is carried by the Higgs boson, referred to as the 'God particle' by Nobel laureate Leon Lederman.

Why does anything have a mass? A truck is heavy because it contains a lot of atoms, each of which might itself be relatively heavy. Steel contains iron atoms and they fall far down the periodic table. But why is an atom heavy? It is mostly empty space after all. Why is a proton heavier than an electron, or a neutrino, or a photon?

Although the four fundamental forces, or interactions, were well known in the 1960s, they all relied on quite different mediating particles. Photons carry information in electromagnetic interactions, gluons link quarks by the strong nuclear force, and the so called W and Z bosons carry weak nuclear forces. But photons have no mass, whereas the W and Z bosons are very massive particles, a hundred times as massive as the proton. Why are they so different? This discrepancy was particularly acute given that the theories of electromagnetic and weak forces could be combined, into an electroweak force. But this theory gave no reason why the weak nuclear force particles, the W and Z bosons, should have a large mass. They should be just like the mass-less photon. Any further combinations of fundamental forces, as attempted by the grand unified theory, also ran into the same problem. Force carriers should not have any mass. Why weren't they all like the photon?

timeline

1687

Newton's *Principia* sets out equations for mass

Slow motion Higgs's big idea was to think of these force carriers as being slowed by passage through a background force field. Now called the Higgs field, it also operates by the transfer of bosons called Higgs bosons. Imagine dropping a bead into a glass. It will take longer to fall to the bottom if the glass is filled with water than if it is empty and filled with air. It is as if the bead is more massive when in water – it takes longer for gravity to pull it through the liquid. The same might apply to your legs if you walk through water – they feel heavy and your motion is slowed. The bead may be slowed even more if dropped into a glass of syrup, taking a while to sink. The Higgs field acts in a similar way, just like a viscous liquid. The Higgs force slows down the other force-carrying particles, effectively giving them a mass. It acts more strongly on the W and Z bosons than on photons, making them appear heavier.

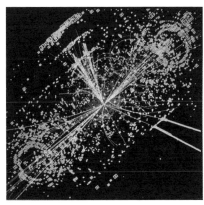

Simulated particle tracks from the decay of a Higgs boson

This Higgs field is quite similar to an electron moving through a crystal lattice of positively charged nuclei, such as a metal. The electron is slowed down a little because it is attracted by all the positive charges, so it appears to be more massive than in the absence of these ions. This is the electromagnetic force in action, mediated by photons. The Higgs field works similarly, but Higgs bosons carry the force. You could also imagine the electron is like a film star walking into a cocktail party full of Higgs's. The star finds it hard to traverse the room because of all the social interactions slowing them down.

If the Higgs field gives the other force-carrier bosons mass, what is the mass of the Higgs boson? And how does it get its own mass? Isn't this a chicken-and-egg situation? Unfortunately physics theories do not predict the mass of the Higgs boson itself, although they do predict the necessity for it within the standard model of particle physics. So physicists expect to see it, but they don't know how hard this will be or when it will appear (it has not been detected yet). Because of the ongoing search for particles with its

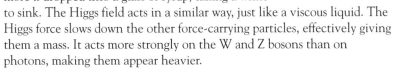

properties, we know that its mass must be greater than the energies already reached experimentally. So it is very heavy, but we must wait to find out exactly how heavy.

Smoking gun The latest machine that will have a good look for the Higgs particle is the Large Hadron Collider (LHC) at CERN in Switzerland. CERN, the Conseil Européen pour la Recherche Nucléaire (European Council for Nuclear Research), is a huge particle physics laboratory near Geneva. It houses rings of tunnels, the largest laid in a circle 27 km long, 100 m below ground. In the LHC, giant magnets accelerate protons forming a beam that curves around the track. They are constantly boosted as they go round, making them race faster and faster. Two opposing beams will be created, and when they are travelling at maximum speed, they will be fired into one another so that the speeding protons smash into each other head on. The huge energies produced will allow a whole range of massive particles to be released temporarily and recorded by detectors, along with their decay products if they are very short-lived.

It is the goal of the LHC to find a hint of the Higgs particle, buried amongst billions of other particle signatures. Physicists know what they are looking for, but it will still be hard to hunt it down. The Higgs may just appear, if the energies are high enough, for a fraction of a second, before disappearing into a cascade of other particles. So, rather than seeing the Higgs itself, the physicists will have to hunt for a smoking gun and then piece everything back together again to deduce its existence.

Symmetry breaking and defects

During the first hundredth of a second after the Big Bang, the universe passed through 4 phases associated with the creation of each of the fundamental forces – electromagnetism, the weak and strong nuclear forces and gravity. Like water condensing from steam to liquid to ice, the universe's structure became more asymmetric as it cooled. Whilst passing through each of these phase changes, imperfections could have arisen, just as ice crystals show flaws in the packing of water molecules within them. Theorists propose that these 'topological defects' in space-time might include linear 'cosmic strings', one-sided magnetic 'monopoles' and twisted forms called 'textures'.

> **The obvious thing to do was to try it out on the simplest gauge theory of all, electrodynamics – to break its symmetry to see what really happens.**
>
> **Peter Higgs**

Symmetry breaking When might a Higgs boson appear? And how do we get from here to photons and other bosons? Because the Higgs boson must be very heavy, it can only appear at extreme energies and, owing to quantum mechanics rules, only then for a very short time indeed. Theories suppose that in the very early universe, all the forces were united together in one superforce. As the universe cooled, the four fundamental forces dropped out, through a process called symmetry breaking.

Although symmetry breaking sounds quite a difficult thing to imagine, in fact it is quite simple. It marks the point where symmetry is removed from a system by one occurrence. An example is a round dinner table set with napkins and cutlery. It is symmetric in that it doesn't matter where you sit; the table looks the same. But if one person picks up their napkin the symmetry is lost – you can tell where you are relative to that position. So symmetry breaking has occurred. Just this one event can have knock-on effects – it may mean that everyone else picks up the napkin to their left, to match the first event. If the first person had taken the napkin from the other side, then the opposite may happen. But the pattern that follows is set up by the random event that triggered it. Similarly, as the universe cooled, events caused the forces to decouple, one by one.

Even if scientists do not detect the Higgs boson with the LHC, it will be an interesting result. From neutrinos to the top quark, there are 14 orders of magnitude of mass that the standard model needs to explain. This is hard to do even with the Higgs boson, which is the missing ingredient. If we do find this God particle, all will be well, but if it is not there then the standard model will need to be fixed. And that will require new physics. We think we know all the particles in the universe – the Higgs boson is the one remaining missing link.

the condensed idea
Swimming against the tide

27 String theory

Even before the standard model has been tested to destruction, or affirmation, some scientists are looking for an alternative view of the stuff of the universe. A group of physicists is trying to explain the patterns of fundamental particles by treating them not as hard spheres but as waves on a string. This idea is known as string theory.

String theorists are not satisfied that fundamental particles, such as quarks, electrons and photons, are indivisible lumps of matter or energy. The patterns that give them a particular mass, charge or associated energy suggest another level of organization. These scientists conceive that such patterns indicate deep harmonies. Each mass or energy quantum is a harmonic tone of the vibration of a tiny string. So particles can be thought of not as solid blobs but as vibrating strips or loops of string. In a way, this is a new take on Kepler's love of ideal geometric solids. It is as if the particles form a pattern of notes that suggest a harmonic scale, played on a single string.

Vibrations The strings in string theory are not as we know them on, say, a guitar. A guitar string vibrates in three dimensions of space, or perhaps we could approximate this to two if we imagine it is restricted to a plane along its length and up and down. But subatomic strings vibrate in just one dimension, rather than the zero dimensions of point-like particles. Their entire extent is not visible to us, but to do the mathematics, the scientists calculate the strings' vibrations over more dimensions, up to 10 or 11 of them. Our own world has three dimensions of space and one more of time. But string theorists think that there may be many more that we don't see, dimensions that are all curled up so we don't notice them. It is in these other worlds that the particle strings vibrate.

timeline

> **'Having those extra dimensions and therefore many ways the string can vibrate in many different directions turns out to be the key to being able to describe all the particles that we see.'**
> **Edward Witten**

The strings may be open-ended or closed loops, but they are otherwise all the same. So the variety in fundamental particles arises only from the pattern of vibration of the string, its harmonics, not from the material of the string itself.

Offbeat idea String theory is an entirely mathematical idea. No one has ever seen a string, and no one has any idea how to know if one were there for sure. So there are no experiments that anyone has yet devised that could test whether the theory is true or not. It is said that there are as many string theories as there are string theorists. This puts the theory in an awkward position among scientists.

The philosopher Karl Popper thought that science proceeds mainly by falsification. You come up with an idea, test it with an experiment and if it is false then that rules something out, so you learn something new and science progresses. If the observation fits the model then you have not learned anything new. String theory is not fully developed, so it does not yet have any definite falsifiable hypotheses. Because there are so many variations of the theory, some scientists argue that it is not real science. Debates about whether it is useful or not fill the letters pages of journals and even newspapers, but string theorists feel that their quest is worthwhile.

Theory of everything By trying to explain the whole zoo of particles and interactions within a single framework, string theory attempts to come close to a 'theory of everything', a single theory that unifies all four fundamental forces (electromagnetism, gravity and the strong and weak

mid 1970	**1984–6**	**1990s**
A quantum gravity theory is obtained	Rapid expansion of string theory explains all particles	Witten and others develop M-theory in 11 dimensions

nuclear forces) and explains particle masses and their properties. It would be a deep theory that underlies everything. In the 1940s, Einstein tried to unify quantum theory and gravity, but he didn't succeed and nor has anyone since. He was derided for his efforts, as unifying the two was thought impossible and a waste of time. String theory brings gravity into the equation, and its potential power draws people to pursue it. However, it is a long way from being precisely formulated, let alone verified.

String theory arose as a novelty, owing to the beauty of its mathematics. In the 1920s Theodor Kaluza used harmonics as a different way to describe some unusual properties of particles. Physicists realized that this same mathematics could describe some quantum phenomena too. Essentially, the wavelike mathematics works well for both quantum mechanics and its extension into particle physics. This was then developed into early string theories. There are many variants, and it remains some way off from an all-encompassing theory.

M-theory

Strings are essentially lines. But in multidimensional space they are a limiting case of geometries that might include sheets and other many-dimensional shapes. This generalized theory is called M-theory. There is no single word that the 'M' stands for, but it could be membrane, or mystery. A particle moving through space scrawls out a line; if the point-like particle is dipped in ink, it traces out a linear path, which we call its world line. A string, say a loop, would trace out a cylinder. So we say it has a world sheet. Where these sheets intersect, and where the strings break and recombine, interactions occur. So M-theory is really a study of the shapes of all these sheets in 11-dimensional space.

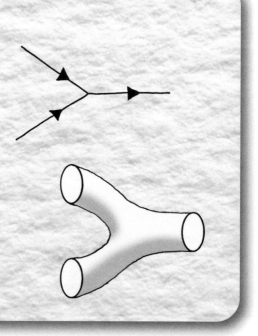

> **❝I don't like that they're not calculating anything. I don't like that they don't check their ideas. I don't like that for anything that disagrees with an experiment, they cook up an explanation – a fix-up to say, "Well, it still might be true."❞**
> **Richard Feynman**

A theory of everything is a goal of some physicists, who are generally reductionists and think that if you understand the building blocks then you can understand the whole world. If you understand an atom, built from vibrating strings, then you can infer all of chemistry, biology and so on. Other scientists find this whole attitude ridiculous. How can a knowledge of atoms tell you about social theory or evolution or taxes? Not everything can be scaled up so simply. They think that such a theory describes the world as a pointless noise of subatomic interactions and is nihilistic and wrong. The reductionist viewpoint ignores evident macroscopic behaviour, such as the patterns of hurricanes or chaos, and is described by physicist Steven Weinberg as 'chilling and impersonal. It has to be accepted as it is, not because we like it, but because that is the way the world works'.

String theory, or rather theories, is still in a state of flux. No final theory has yet emerged, and this may take some time, as physics has become so complicated that there is a lot to include in it. Seeing the universe as the ringing of many harmonies has charm, but its adherents also sometimes verge on the dry side, being so engrossed in the fine detail that they diminish the significance of larger-scale patterns. Thus string theorists may stay on the sidelines until a stronger vision emerges. But given the nature of science, it is good that they are looking, and not in the usual places.

the condensed idea
Universal harmonies

28 Anthropic principle

The anthropic principle states that the universe is as it is because if it were different we would not be here to observe it. It is one explanation for why every parameter in physics takes the value that it does, from the size of the nuclear forces to dark energy and the mass of the electron. If any one of those varied even slightly, then the universe would be uninhabitable.

If the strong nuclear force was a little different, then protons and neutrons would not stick together to make nuclei and atoms could not form. Chemistry would not exist. Carbon would not exist, and so biology and humans would not exist. If we did not exist, who would 'observe' the universe and prevent it from being only a quantum soup of probability?

Equally, even if atoms existed and the universe had evolved to make all the structures we know today, then if dark energy were a little stronger, galaxies and stars would already be being pulled apart. So, tiny changes in the values of the physical constants, in the sizes of forces or masses of particles, can have catastrophic implications. Put another way, the universe appears to be fine-tuned. The forces are all 'just right' for humanity to have evolved now. Is it a chance happening that we are living in a universe that is 14 billion years old, where dark energy and gravity balance each other out and the subatomic particles take the forms they do?

Just so Rather than feel that humanity is particularly special and the entire universe was put in place just for us, perhaps a rather arrogant assumption, the anthropic principle explains that it is no surprise. If any of

Alfred Wallace discusses
man's place in the universe

❝In order to make an apple pie from scratch, you must first create the universe.❞

Carl Sagan

the forces were slightly different, then we simply would not be here to witness it. In the same way that there are many planets but as far as we know only one that has the right conditions for life, the universe could have been made in many ways but it is only in this one that we would come to exist. Equally, if the combustion engine had not been invented when it was and my father had not been able to travel north to meet my mother, then I would not be here. That does not mean that the entire universe evolved thus just so that I could exist. But the fact that I exist ultimately requires, amongst other things, that the engine was invented beforehand, and so narrows the range of universes that I might be found in.

The anthropic principle was used as an argument in physics and cosmology by Robert Dicke and Brandon Carter, although its theory is familiar to philosophers. One formulation, the weak anthropic principle, states that we would not be here if the parameters were different, so the fact that we exist restricts the properties of inhabitable physical universes that we could find ourselves in. Another stronger version emphasizes the importance of our own existence, such that life is a necessary outcome for the universe coming into being. For example, observers are needed to make a quantum universe concrete by observing it. John Barrow and Frank Tipler suggested yet another version, whereby information processing is a fundamental purpose of the universe and so its existence must produce creatures able to process information.

Many worlds To produce humans, you need the universe to be old, so that there's enough time for carbon to be made in earlier generations of stars, and the strong and weak nuclear forces must be 'just so' to allow nuclear physics and chemistry. Gravity and dark energy must also be in balance to make stars rather than rip apart the universe. Further, stars need

1957
Robert Dicke writes that the universe
is constrained by biological factors

1973
Brandon Carter discusses
the anthropic principle

Anthropic bubbles

We can avoid the anthropic dilemma if many parallel, or bubble, universes accompany the one we live in. Each bubble universe can take on slightly different parameters of physics. These govern how each universe evolves and whether a given one provides a nice niche in which life can form. As far as we know, life is fussy and so will choose only a few universes. But since there are so many bubble universes, this is a possibility and so our existence is not so improbable.

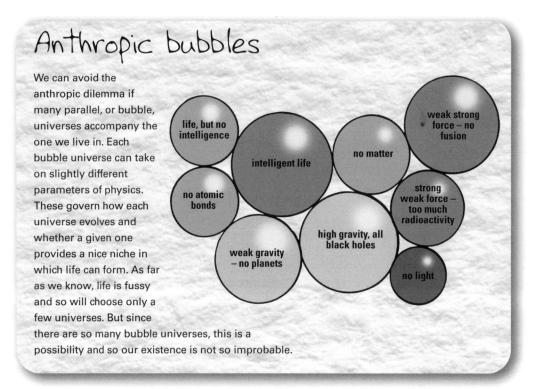

to be long-lived to let planets form, and large enough so that we can find ourselves on a nice suburban temperate planet that has water, nitrogen, oxygen and all the other molecules needed to seed life.

Because physicists can imagine universes where these quantities are different, some have suggested that those universes can be created just as readily as one like our own. They may exist as parallel universes, or multi-verses, such that we only exist in one realization. The idea of parallel universes fits in with the anthropic principle in allowing other universes to exist where we cannot. These may exist in multiple dimensions and are split off when any event occurs.

On the other hand The anthropic principle has its critics. Some think it is a truism – it is like this because it is like this – and is not telling us much that's new. Others are unhappy that we have just this one special universe to test, and prefer to search the mathematics for ways of automatically tuning

> **The observed values of all physical and cosmological quantities are not equally probable but they take on values restricted by the requirement that there exist sites where carbon-based life can evolve and . . . that the Universe be old enough for it to have already done so.**
> John Barrow and Frank Tipler

our universe to fall out of the equations simply because of the physics. The multiverse idea comes close to this by allowing an infinite number of alternatives. Yet other theorists, including string theorists and proponents of M-theory, are trying to go beyond the Big Bang to fine-tune the parameters. They look at the quantum sea that preceded the Big Bang as a sort of energy landscape and ask where a universe is most likely to end up if you let it roll and unfold. For instance, if you roll a ball down a ridged hill, then the ball is more likely to end up in some places than others, such as in valley floors. So in trying to minimize its energy, the universe may well seek out certain combinations of parameters quite naturally, irrespective of whether we are a product of it billions of years later.

Proponents of the anthropic principle, and others who pursue more mathematical means of ending up with the universe we know, disagree about how we got to be where we are and whether that is even an interesting question to ask. Once we get beyond the Big Bang and the observable universe, into the realms of parallel universes and pre-existing energy fields, we are really on philosophical ground. But whatever triggered the universe to appear in its current garb, we are lucky that it has turned out this way billions of years hence. It is understandable that it takes time to cook up the chemistry needed for life. But why we should be living here at a particular time in the universe's history, when dark energy is relatively benign and balancing out gravity, is more than lucky.

the condensed idea
The just-so universe

29 Hubble galaxy sequence

Galaxies come in two types – elliptical and spiral. Astronomers have long suspected that similarities and differences between them, such as their common central bulges and the presence or absence of a flattened disc of stars, indicate an evolving trend. Evidence that galaxy collisions might be responsible for this 'Hubble sequence', stems from the deepest images taken of the sky.

Once it became accepted in the 1920s that some of the fuzzy nebulae that spatter the heavens were galaxies beyond our own, astronomers sought to categorize them. Galaxies fall into two basic types – some are smooth and ellipsoidal in shape; others have clear spiral patterns superimposed on them. These classes are known as elliptical and spiral galaxies, respectively.

Edwin Hubble, the American astronomer who was the first to establish that the nebulae lay outside the Milky Way at vast distances, proposed that galaxies formed a sequence and named them accordingly. His classifications are still in use today. Elliptical galaxies are described by the letter E followed by a number (from 0 to 7) that increases according to how elongated the galaxy is. An E0 galaxy is roughly round; an E7-type galaxy is more cigar-shaped. In three dimensions, ellipticals are shaped like an American football (or rugby ball).

Spiral galaxies, in Hubble's scheme, are accorded the letter S and an extra letter (a, b or c) depending on the degree to which their spiral arms are tightly wound. An Sa galaxy is a tight spiral; one classed as Sc forms a loose

timeline

spiral. In three dimensions, spiral galaxies are flattened like a (solid) frisbee disc or lens. A complication is that some spiral galaxies have a linear feature, or 'bar', across the inner regions. These galaxies are called barred spirals, and follow the same naming pattern, though labelled SB rather than S. Galaxies fitting neither scheme include those of irregular shape, called irregulars, and also galaxies that lie somewhere between ellipticals and spirals, which are classed as S0.

Hubble's tuning fork

If you look more closely, there are similarities between the structures of the two classes. Spirals are formed of two components, like a fried egg: a central bulge (the yolk), which looks a lot like an elliptical galaxy, and the flat disc (the white) that skirts it. The size of the bulge relative to the disc is another means of categorizing galaxies. Hubble even imagined that galaxies formed a sequence, from ones dominated by the bulge, including ellipticals, to those that are almost entirely discs. The former are sometimes referred to as 'early' types; and the latter 'late'-type galaxies. Hubble thought that these similarities meant that galaxies could evolve from one type to another.

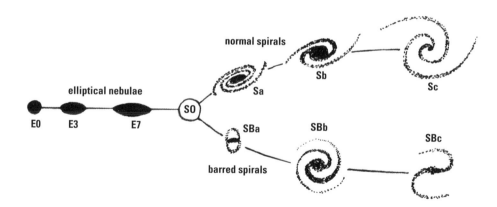

Hubble arranged his categories of galaxy on a tuning-fork-shaped diagram. From left to right along the handle of the fork he drew a sequence of elliptical galaxies, from rounded to elongated. Tacked on to the right, along the upper prong, was a sequence of spirals running from tight-wound spirals with large bulges and small discs through to discs with loose spirals and hardly any bulge. Barred spirals ran along a parallel prong beneath. In this famous diagram – called the Hubble Tuning Fork (se page 117) – Hubble expressed the powerful idea that ellipticals could grow discs and one day become spirals. However, he had no evidence that such transformations happened. Many researchers since have spent their entire careers trying to figure out how galaxies might evolve from one type to another.

Mergers One way in which galaxies can change their character radically is through collisions. Whilst mapping the sky through telescopes, astronomers have found many close pairs of galaxies that are clearly interacting. In the most dramatic cases, long tadpole-like tails of stars are dragged out of both galaxies by their mutual gravity, such as in the pair of colliding galaxies called the 'Antennae' galaxies. Other galaxies have ploughed straight through the middle of a companion, punching out clouds of stars and shedding smoke rings of gas. The ensuing disturbance often makes the galaxies shine extra brightly, as new stars are formed in turbulent gas clouds. These young blue stars may be enshrouded in cosmic soot, making regions glow red, in the same way that dust enhances a sunset. Merging galaxies can be spectacular.

Yet the details of how galaxies are built remain uncertain. It would take a catastrophic collision to destroy a large disc of stars and leave a naked elliptical bulge; or a series of gentle accumulations to let a galaxy gently grow a sizeable disc without disruption. Astronomers see few galaxies in states in between, so the true picture of how galaxies change through mergers is likely to be complicated.

Galactic ingredients Galaxies contain from millions to trillions of stars. Elliptical galaxies, and the bulges of spirals, contain mostly old red stars. These travel on randomly inclined orbits generating their puffed-up ellipsoidal shape. The discs of spiral galaxies, in contrast, display mainly young blue stars. These concentrate in the spiral arms, which trigger star formation as they sweep through gas clouds in the disc. The discs of spirals contain a lot of gas, especially hydrogen. In contrast, elliptical galaxies host little gas and so fewer new stars are formed in them.

> **At the last dim horizon, we search among ghostly errors of observations for landmarks that are scarcely more substantial. The search will continue. The urge is older than history. It is not satisfied and it will not be oppressed.**

Edwin Hubble

It was in galactic discs too that dark matter was discovered (see p.72). The outskirts of spirals rotate too quickly to be explained only by their mass in stars and gas, implying that some other form of matter is present. That extra material isn't visible – it doesn't emit or absorb light – and is called dark matter. It might be in the form of exotic particles that are difficult to detect because they rarely interact, or compact weighty objects such as black holes, failed stars or gas planets. Dark matter forms a spherical cocoon around a galaxy, which is referred to as its 'halo'.

Hubble deep field The same basic types of galaxies exist right across the universe. The deepest image of the sky yet taken is in the Hubble Deep Field. To see what an average swath of the distant universe looks like, in 1995 the Hubble Space Telescope observed a small patch of the sky (2.5 arcminutes across) over a period of 10 days. The orbiting observatory's sharp eyesight meant that astronomers could see much deeper than was possible with ground-based telescopes, and a vista of distant galaxies was opened up. Because light takes time to reach us across the expanse of space, these galaxies are seen as they were many billions of years ago.

Because the field was chosen deliberately to be clear of foreground stars, nearly all of the 3,000 objects in the frame are far-off galaxies. The majority are identifiable as ellipticals and spirals, indicating that both types were formed long ago. But more irregulars and small blue galaxies appear in the distant universe than nearby. Also, stars were being formed at ten times the rate 8–10 billion years ago that they are today. Both these factors suggest that more frequent collisions are responsible for the rapid growth of galaxies in the young universe.

the condensed idea
Galactic transformers

30 Galaxy clusters

Galaxies group together to form clusters, which are the biggest objects in the universe that are bound by gravity. As massive accumulations of thousands of galaxies, clusters also collect reservoirs of super-hot gas and dark matter, which are spread out in between the cluster members.

In the 18th century, astronomers realized that the nebulae were not evenly spread. Just like the stars, they often huddled in groups and clusters. French astronomer Charles Messier was one of the first to survey and list the brightest nebulae – including what we now know to be galaxies, as well as diffuse and planetary nebulae, star clusters and globular clusters. The first version of his catalogue, published in 1774 in the journal of the French Academy of Sciences, included just 45 of the most spectacular smudges; a later version in 1781 listed over a hundred. Astronomers still name Messier's objects with the letter M prefix and a catalogue number – the Andromeda galaxy, for instance, is also known as M31. Messier's catalogue includes some of the most closely studied objects of their class.

A much larger catalogue of deep sky objects – the New General Catalogue – was compiled and published in the 1880s. In it Johann Dreyer listed almost 8,000 objects, of which nearly a third were from observations by William Herschel. Different object types were distinguished in various classes, from bright nebulae to loose star clusters. After the advent of photography made it possible to find many more objects, the catalogue was expanded in 1905 with the addition of two Index Catalogues comprising more than 5,000 objects. These astronomical objects are still named NGC or IC depending on which catalogue they were listed in. The Andromeda galaxy, for example, is also referred to as NGC 224.

timeline

1781
Messier notes Virgo cluster

1924
Hubble measures distance to Andromeda galaxy

Who are we? We find that we live on an insignificant planet of a humdrum star lost in a galaxy tucked away in some forgotten corner of a universe in which there are far more galaxies than people.

Carl Sagan

The Local Group In the 1920s, astronomers found that many nebulae were galaxies remote from our own. Using the cosmic distance ladder techniques, including Cepheid variable stars and redshifts, they could estimate their distances: the Andromeda galaxy, for instance, is 2,500,000 light years away. It soon became clear that Andromeda and the Milky Way are the two largest members of a group of about 30 galaxies, known as the Local Group.

Andromeda and the Milky Way are quite similar in size and character. The Andromeda galaxy is also a large spiral, although we view it on its side, inclined by about 45 degrees. The other galaxies in the group are much smaller. Our two nearest neighbours, some 160,000 light years away, are the Large and Small Magellanic Clouds, which appear as thumb-sized smudges in the southern sky adjacent to the Milky Way band. They are named for explorer Ferdinand Magellan, who brought back reports of them after he circumnavigated the globe in the 16th century. The Magellanic Clouds are irregular dwarf galaxies about a tenth the size of the Milky Way.

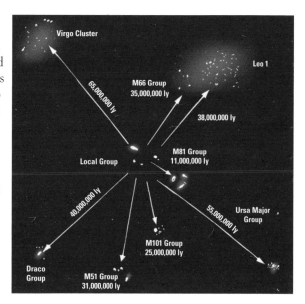

Virgo Cluster

Leo 1

65,000,000 ly

M66 Group
35,000,000 ly

38,000,000 ly

M81 Group
11,000,000 ly

Local Group

40,000,000 ly

Ursa Major
Group

55,000,000 ly

M101 Group
25,000,000 ly

Draco
Group

M51 Group
31,000,000 ly

1933

Zwicky measures dark
matter in comacluster

1966

X-rays detected from
Virgo cluster

Virgo cluster The Local Group is one of many collections of galaxies. The much richer Virgo cluster contains thousands of galaxies, of which 16 were bright enough to be noted as grouped together in Messier's catalogue of 1781. Virgo is the nearest large galaxy cluster to us, some 60 million light years away. Other examples of vast clusters include the Coma cluster and the Fornax cluster, each named for the constellation in which it sits. In fact, the Virgo cluster and the Local Group are part of an even larger concentration called the Local Supercluster.

Galaxy clusters are held together by gravity. Just as stars follow orbits within galaxies, so galaxies trace out paths around the cluster's centre of mass. A typical large galaxy cluster has a total mass 10^{15} (a million billion) times that of the Sun. Moreover, by packing so much matter into a small space, spacetime itself distorts. Using a rubber sheet analogy, the weight of the galaxies presses down so that they all sit in a depression. But it is not just galaxies that fall in – gas also accumulates in the spacetime pit.

Intra-cluster medium Galaxy clusters are full of hot gas. Because it is so hot – millions of degrees Celsius – this pool of gas glows brightly enough to give off X-rays, which can be detected with satellites. The hot gas is referred to as the intra-cluster medium. In a similar way, dark matter also gathers in the gravitational well of clusters. Because astronomers hope to see the dark matter in a new environment outside individual galaxies, they

CHARLES MESSIER (1730–1817)

Messier was born into a large family in the Lorraine region of France. He became interested in astronomy after a spectacular six-tailed comet appeared in the sky in 1744, followed by a solar eclipse that he would have witnessed in his home town in 1748. In 1751 he joined the navy as an astronomer, documenting carefully observations such as the transit of Mercury across the Sun's face in 1753. He was widely recognized by European scientific institutions, and in 1770 was elected to the French Academy of Sciences. Messier created a famous catalogue, in part to help comet-hunters of the day. He discovered 13 comets and has a crater on the moon and an asteroid named after him.

> **The image is more than an idea. It is a vortex or cluster of fused ideas and is endowed with energy.**
>
> **Ezra Pound**

are looking within clusters for unusual signs that might help them understand what dark matter is made of. One study, for instance, has claimed to find a speeding 'bullet' of dark matter moving differently from the hot gas that surrounds it in a particular cluster. But still dark matter's origin is mysterious. Because clusters are so massive, they can also distort the light from galaxies lying behind them. Bending the light as it passes, they act as giant but grainy 'gravitational lenses' (see p.148), smearing out distant galaxies into curves and smudges.

Clusters can be thought of unflatteringly as the garbage heaps of the cosmos, because they are so big that everything falls into them. They are therefore intriguing places for cosmic archaeologists. Moreover, as the largest gravitationally bound objects, they should contain proportions of normal and dark matter that are representative of the universe as a whole. If we can count and weigh all the clusters, then that approximates to the total mass of the universe. And if we can track how they grow over time, by looking for very distant clusters seen just as they might be forming, then we can learn how the structure of the universe has developed since the Big Bang.

Coma cluster

the condensed idea
Everything collects here

31 Large-scale structure

Galaxies are spread across the universe in foam-like structures. Clusters lie at the intersections of filaments and sheets, which wrap around empty regions called voids. This cosmic web is the result of billions of years of gravity, pulling galaxies towards one another since their birth.

By the 1980s, astronomers' instruments had become so advanced that they could measure redshifts for many galaxies simultaneously by recording their light characteristics as multiple spectra. A group of astronomers from Harvard's Center for Astrophysics (CfA) decided to systematically collect redshifts for hundreds of galaxies, to try to reconstruct their positions in space in three dimensions. Their resulting survey, known as the CfA Redshift Survey, revealed a new view of the cosmos.

The astronomers mapped out the Milky Way's neighbourhood, from its Local Group out to the nearest clusters and the supercluster on whose edge we lie. As the survey grew, it probed further. In 1985, the astronomers had collected over a thousand redshifts, reaching out to 700 million light years away. By 1995, the survey had claimed over 18,000 redshifts for relatively bright galaxies over a wide area in the northern sky.

Cosmic foam The first map was surprising. It showed that even on these huge scales, the universe was not random. The galaxies were not evenly spread but seemed to cling to invisible filaments, strung out in arcs on the surfaces of bubbles around empty regions called voids. This foam-like structure is known as the 'cosmic web'. Clusters of galaxies formed

timeline

1977	1985
CfA Redshift Survey begins	Great Wall of galaxies discovered

> **You can't construct a doctrine of creation without taking account of the age of the universe and the evolutionary character of cosmic history.**
> John Polkinghorne

where filaments overlapped. The biggest structure in the survey was dubbed The Great Wall – a band of galaxies concentrated in a vast region of dimensions 600 by 250 by 30 million light years. Embedded in this swath are many galaxy clusters, including the famous Coma cluster, one of the most massive near us.

Since the first surveys, technology has made it even simpler to collect redshifts, and today's attempts have mapped out millions of galaxies across most of the sky. The largest is the Sloan Digital Sky Survey, whose intensive observations are made year after year from a dedicated 2.5 m diameter telescope at Apache Point Observatory in New Mexico. Started in 2000, the survey aims to image 100 million objects over 25 per cent of the sky and to get redshifts for a million of those. It does so by grabbing 640 spectra at a time, using optical fibres that are stuck on holes drilled into a metal plate. Every patch of the sky has to have its own plate specially made; up to nine plates are used each night.

Galaxy segregation The Sloan survey has given us a pristine view of the structures of galaxies in the universe. At every scale measured, galaxies follow similar web-like patterns. Because the survey collects spectra and also images, astronomers can distinguish different types of galaxies. Elliptical galaxies tend to be relatively red and their spectra are similar to the light from old stars. Spiral galaxies are bluer and their spectra reveal younger stars, as they form in their gas-rich discs.

The Sloan survey reveals that different types of galaxies congregate in different ways. Elliptical galaxies favour clusters and crowded regions of space. Spirals are more widely scattered, and dislike the centres of rich

2000
Sloan Digital Sky Survey begins

2015
Large Synoptic Survey Telescope comes online

galaxy clusters. Although by definition they are mostly empty, voids can contain a smattering of galaxies, usually spirals. This segregation indicates that galaxies know about their environments.

Quasar absorption lines   Although glowing galaxies are easy to trace, less is known about how dark matter and gas are spread through space. Gas clouds can be spotted when they absorb the light from objects lying behind them. Quasars, being very bright objects found typically very far away, make good lighthouses against which to search. Just as it absorbs sunlight, making Fraunhofer spectral lines (see p.28), hydrogen gas leaves recognizable signatures in the light spectrum of quasars. So clouds of hydrogen gas can be located through the absorption lines they produce. Other trace elements within the cloud can also be measured, although those absorption lines are often weaker and harder to spot.

The strongest absorption line of hydrogen appears in the ultraviolet region of the spectrum (at wavelength 121.6 nanometers); when it is redshifted it then appears at longer wavelengths. It is called the Lyman-alpha line. Hydrogen-rich gas clouds, often little polluted since the Big Bang, which produce this absorption line are sometimes called Lyman-alpha clouds. If there are many clouds in front of the quasar light source, then each one will produce a gap in the spectrum at a wavelength that corresponds to its redshift. The resulting series of black lines cut in to the ultraviolet light emitted by the quasar is called the Lyman-alpha forest.

If sightlines to many background quasars are probed, then distributions of hydrogen gas clouds in front of them can be estimated. On the whole, astronomers see that the gas also follows closely the structures picked out

Future surveys

The next generations of surveys hope to take movie-like sequences of the entire sky in multiple colours. The Large Synoptic Survey Telescope is an 8.4 m diameter telescope with a three-billion-pixel digital camera, which is currently being built in Chile. Covering 49 times the area of the Moon in a single exposure, from 2015 it will be able to image the sky every week. Such telescopes will probe the mysteries of dark matter and dark energy and detect objects that change or move, such as supernovae and asteroids.

by galaxies. Less is known about dark matter, because it doesn't interact with light so cannot be seen glowing or in absorption. But astronomers suspect that it too favours the galactic conurbations.

Gravity's pull The cosmic web is ultimately caused by gravity, operating on galaxies since they formed. Out of the primordial hydrogen that infused the early universe after the Big Bang grew stars and galaxies. The galaxies were then drawn together over time, such that the filaments, clusters and walls developed.

Astronomers know roughly how matter was distributed 400,000 years ago, because that is when the cosmic microwave background radiation was released. The hot and cold spots in it tell us how lumpy the universe was then. The redshift surveys tell us how lumpy it is today, and in the recent past. Astronomers then try to tie together the two snapshots – like trying to work out how a baby grew up to be an old man, they work out what processes turned the universe from its infant stage to maturity.

The precise pattern of the cosmic foam depends sensitively on many parameters in cosmological theories. By tweaking them, astronomers can constrain the geometry of the universe, the amount of stuff in it, and also the characteristics of the dark matter and dark energy. To do this they run vast computer simulations, putting all their data (galaxies, gas and dark matter) into them, and cranking the handle to estimate the parameters.

Even so, the answer isn't simple. The nature of the dark matter makes a big difference, and we simply don't know what it is. Models that consider 'cold' types of dark matter – slow-moving exotic particles – predict stronger clustering on large scales than is seen. If the dark matter particles are fast-moving, that is, 'hot' or 'warm', then they would smear out fine-scale structure more than is seen also. So the galaxy clustering data suggest that dark matter lies somewhere in between. Similarly, too much dark energy counteracts gravity and slows the accumulation of galaxies. The best bet universe is a compromise.

> **'Fiction is like a spider's web, attached ever so slightly perhaps, but still attached to life at all four corners. Often the attachment is scarcely perceptible.'**
>
> **Virginia Woolf**

the condensed idea
Cosmic web

32 Radio astronomy

Radio waves open up a new window on the violent universe. Produced by exploding stars and jets emanating from black holes, radio waves identify fast-moving particles in strong magnetic fields. Their most extreme examples are radio galaxies, where twin jets fuel bubble-like lobes far beyond the galaxy's stars. The distribution of radio galaxies also backs up the Big Bang model of the universe.

Cosmic microwave background radiation wasn't the only astronomical discovery to come from trying to account for static in radio receivers (see p.60). In the 1930s, an engineer with Bell Telephone Laboratories called Karl Jansky was investigating noise that troubled his short-wave transatlantic voice transmissions. He found a signal that chipped in every 24 hours. At first he suspected it might be the Sun – as other scientists, including Nikola Tesla and Max Planck, had predicted that our star should emit electromagnetic waves right across the spectrum. Listening for longer, he found that it wasn't coming from that direction. Its timing was also slightly less than 24 hours; it matched the daily rotation of the sky as seen from the spinning Earth, implying that it had a celestial origin.

In 1933, Jansky worked out that the static was coming from the Milky Way, mostly from the constellation Sagittarius, which hosts our galaxy's centre. The fact that it wasn't coming from the Sun indicated that it must arise not in stars but in interstellar gas and dust. Jansky didn't stay in astronomy. Nevertheless he is remembered as the father of radio astronomy, and a unit of radio brightness (flux density) was named the Jansky (Jy) after him.

timeline

1933	1937
Jansky detects Milky Way in radio	Reber builds first radio telescope

Another pioneer was Grote Reber, a ham radio enthusiast from Chicago, Illinois, who built the first radio telescope in his back yard in 1937. He built a parabolic reflecting dish over 30 feet in diameter and secured a radio detector at its focus, 20 feet above. The radio receiver amplified the cosmic radio waves millions of times. These electronic signals were then fed to a pen plotter and recorded as charts.

Radio telescopes Although today's radio telescopes can operate during the day (they are unaffected by sunlight), Reber observed at night in order to avoid contamination from sparks from automobile engines. During the 1940s, he surveyed the sky in radio waves. Plotting its brightness as a contoured map, he sketched out the shape of the Milky Way, with the brightest emissions coming from the galaxy's centre. He also detected several other bright sources of radio waves, including ones in the constellations of Cygnus and Cassiopeia. It was not until 1942 that British army research officer J.S. Hey detected radio waves from the Sun.

New Radio Waves Trace to Center of the Milky Way ... No Evidence of Interstellar Signalling.

New York Times, 1933

Although the science of radio astronomy took off after the Second World War, much of the technology arose as countries raced to build radar systems. Radar – short for RAdio Detection And Ranging – also led to many electronic devices that made possible much of the technology we use today.

Surveys By the early 1950s, physicists in the UK and Australia were producing surveys of the radio sky using a technique called radio interferometry. Whereas Reber's telescope used a single dish and detector, like a mirror in an optical reflecting telescope, radio interferometers use many detectors spread out over a wider distance. The spread is equivalent

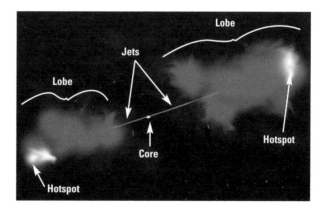

1953

Cygnus A shown to be
double radio source

1959

3C radio survey published

Cosmic hiss

You can detect the hiss from the Milky Way if you have a small radio set. Tune it away from any stations, so that you just hear the static. Then wave its antenna around and you will find that the hiss gets louder and softer. The extra hiss is due to it picking up radio waves from the Milky Way.

to using a big mirror; but by combining the signals from many detectors, astronomers can image regions of the sky in finer detail than is possible with one very large dish. Such a set-up is ideal for making surveys.

Using a radio interferometer in Cambridge, British physicists Antony Hewish and Martin Ryle began a series of surveys of the brightest radio sources in the northern sky, working at a frequency of 159 MHz. Following two prior listings, their resulting 3rd Cambridge Survey, or 3C for short, the first of suitably high quality, was published in 1959. The earlier versions were beset by calibration problems, and their veracity led to clashes with astronomers in Australia who were making surveys of the southern sky in parallel. Between 1954 and 1957, Bernard Mills, Eric Hill and Bruce Slee, using the Mills Cross telescope in New South Wales, recorded and published lists of over 2,000 radio sources. By the time the 3C was published, differences between the researchers were resolved, and the radio sky was open to investigation in both hemispheres.

The nature of the radio sources was the next question – and optical spectra were sought. However, because the radio source positions were only crudely known, it was hard to work out which star or galaxy was responsible. Eventually, the sources yielded their secrets. Apart from the Milky Way's centre, some of the brightest sources are unusual objects in our galaxy. For example, Cassiopeia A and the Crab nebula are supernova remnants, shells of gas blown out by the catastrophic explosion of a dying star, the latter with a pulsar in its centre.

Radio galaxies Other sources are more extreme. The bright source in Cygnus – known as Cygnus A – is a distant galaxy. Discovered by Reber in 1939, it was shown in 1953 to be not one source but two. Such double sources are characteristic of many radio-emitting galaxies. Either side of the galaxy are two diffuse 'lobes', vast bubbles inflated by fine jets of energetic particles that emanate from the galaxy's centre. The symmetry of the lobes – they are usually equidistant and of similar sizes and shapes – suggests they are fuelled by a single engine. That engine is thought to be a black hole

> **[The Big Bang] is an irrational process that cannot be described in scientific terms ... [nor] challenged by an appeal to observation.**
>
> Fred Hoyle

that lurks in the centre of the radio galaxy. As material is sucked towards the black hole, it gets ripped apart into its constituent particles, which are blown out by the jets at speeds approaching that of light. Radio waves arise because the particles interact with strong magnetic fields to produce 'synchrotron radiation'. Most radio waves in space come from interactions between particles and magnetic fields – in the hot diffuse gas that infuses our own galaxy and galaxy clusters, in jets or near compact objects where magnetic fields are intensified, such as black holes. The centre of the Milky Way too hosts a black hole.

Ryle v Hoyle The number of radio sources in the universe proved to be critical for the Big Bang theory. Ryle, strident leader of the radio astronomy group at Cambridge University, famously confronted Fred Hoyle, a charismatic astronomer across the road at the Institute of Astronomy, who had worked on the process of nucleosynthesis, the creation of elements in stars and the Big Bang. In the days before the discovery of the cosmic microwave background, the Big Bang model wasn't accepted – in fact it was Hoyle himself who coined the phrase 'Big Bang' as ridicule. He favoured instead a 'steady state' picture of the universe, arguing that it did not have a beginning and had always existed. So he expected that galaxies would be scattered through space at random, stretching on infinitely.

But Ryle had found evidence that radio sources were not uniformly spread – he saw that there were more moderately bright radio sources than expected from a random distribution. Thus, he argued, the universe must be finite and the Big Bang model true. Ryle was proved right by the discovery of the cosmic microwave background, although the two great astronomers continued to spar. To this day the two research groups are fiercely independent because of this history of hostility.

the condensed idea
Radio landscape

33 Quasars

Quasars include the most distant and luminous objects in the universe. Their extreme brightness is a result of matter falling towards a central black hole within a galaxy. Because of their geometry they look very different when seen from different directions and can appear as unusual 'active galaxies' with narrow emission lines. All galaxies may go through a quasar phase, which may play an important role in their creation.

During the 1960s, a weird class of star puzzled astronomers. Their unusual spectra showed bright emission lines, but the lines did not seem to lie at the right wavelengths to be attributed to known elements. What were these objects? In 1965, a Dutch astronomer, Maarten Schmidt, realized that the lines did correspond to normal elements, including the characteristic sequence due to hydrogen, but they were heavily redshifted.

The redshifts indicated that these 'stars' lay at vast distances, well beyond the Milky Way and in the realm of the galaxies. Yet they did not look like fuzzy galaxies – they were point sources of light. Furthermore, for the distances that their redshifts indicated, they were exceedingly bright. It was surprising that something that looked like one of the stars in our galaxy was in fact located well beyond the Local Supercluster. What could power such a thing?

Quasars Astronomers realized that the only way to release enough energy to power these extragalactic objects – dubbed 'quasi-stellar objects' or QSOs – was through the extreme action of gravity, namely near black holes. Matter falling towards a black hole in the centre of a galaxy could heat up through friction and radiate enough light to explain the enormous

> **❝If a car was as fuel-efficient as these black holes, it could theoretically travel over a billion miles on a gallon of gas.❞**
> ### Christopher Reynolds

luminosity of the QSOs. The light from the central point outshines the rest of the galaxy, making it appear from afar like a star. A fraction of QSOs, around 10 per cent, also emit radio waves: these are named 'quasi-stellar radio sources', or 'quasars' for short. The whole class is often referred to simply as quasars.

As gas, dust or even stars spiral in towards a black hole, the material collects in a disc, called an 'accretion disc' following Kepler's Laws. Just like the planets in our own solar system, material in the inner parts of the disc orbits more quickly than the outer parts. Adjacent shells of gas rub against each other and heat up to millions of degrees, eventually starting to glow. Astronomers predict that the inner parts of the accretion disc are hot so they emit X-rays; the outer parts are cooler, and emit infrared radiation. Visible light comes from regions in between.

This range of temperatures generates emission over a wide range of frequencies, each temperature corresponding to a characteristic black body spectrum that peaks at a different energy. Thus quasars radiate from the far infrared through to X-rays, a much more extreme range than any star. If strong magnetic fields and particle jets are present, as for radio galaxies, then the quasar also shows radio emission. The presence of such a bright and energetic source of light produces another component that is characteristic of quasars – broad emission lines. Clouds of gas floating just above the disc can be illuminated, causing them to glow in spectral lines that reflect the chemical make-up of the clouds. Because of the proximity of the central black hole, the clouds move very

Quasar environments

Active galactic nuclei may be found in both elliptical and spiral hosts. Yet certain AGN classes favour certain environments. Powerful radio sources tend to be associated with large elliptical galaxies. Spiral galaxies with active nuclei tend to weak radio emitters. Active galaxies are commonly found in galaxy groups and clusters. This has suggested to some scientists that collisions might be involved in triggering black holes into action. If one of the galaxies in a merger is a spiral, then it would bring gas fuel with it that would be funnelled on to the black hole, causing it to flare up.

fast, so these emission lines become broadened due to the Doppler effect. The emission lines in quasars are much broader than in other types of galaxies, where they are typically narrow.

Active galaxies Quasars are the most extreme examples of a class of galaxies with accreting black holes, called active galactic nuclei, or AGN. The presence of black holes is revealed by characteristic emission lines that are difficult to excite except in the highly ionized gas that arises due to the hot temperatures generated near the black hole. Broad lines are visible to us only if the regions nearest the black hole can be seen directly. In other types of AGN the inner regions may be hidden by dense clouds of gas and dust distributed in a doughnut torus shape, and the broad lines obscured. Although only narrow lines remain visible, the high-ionization levels of the lines give away the presence of the monster at the AGN's heart.

These different classes of quasars and other active galaxies may arise simply because we are viewing them from different directions. Many galaxies have more obscuring material around their fattest axis, for instance they may show dust lanes. So if such galaxies were viewed edge on, then this extra material and any central dust torus might block the sightline to any central black hole. Along the shortest axis of the galaxy, a clearer view to the centre is possible. Thus quasars might be predominantly viewed down the short axis; and AGN lacking broad lines from the side. Material might also be cleared out along the shorter axis more easily by outflows; so cones may be opened up, making the view even cleaner.

Unified schemes This idea that different types of AGN might arise simply from viewing direction is known as the 'unified scheme'. The basic idea works quite well for quasars and other active galaxies that are well matched in their large scale properties, such as radio or galaxy brightness. However, AGN come in many variants. The intrinsic brightness of the AGN, through the size of its black hole, is likely to affect how clear the view is to the centre. The centres of weak AGN might be more buried than more powerful ones. Or young AGN, whose central black holes have only recently switched on, might be more heavily obscured than older ones, which have had more time to clear out material. The presence or absence of radio emission is another factor that is unexplained – some astronomers think that radio jets arise from spinning black holes, or following certain types of galaxy collisions, such as between two massive elliptical galaxies.

> **Twinkle, twinkle, quasi-star**
> **Biggest puzzle from afar**
> **How unlike the other ones**
> **Brighter than a billion suns**
> **Twinkle, twinkle, quasi-star**
> **How I wonder what you are.**
>
> **George Gamow**

Feedback Astronomers are gradually learning about how the presence of an accreting black hole in the centre affects the way a galaxy develops. When active, the central black hole may blast gas out of the galaxy, leaving behind less fuel for new stars to form. This might explain, for example, why elliptical galaxies contain little gas and few young stars. In contrast, if AGN switch on after collisions, then any gas introduced may also kick off a rapid burst of star formation – so a galaxy may go through a phase where it is heavily obscured and is building new stars. As the AGN gets going, it clears away the mess and blasts out the extraneous gas, but this then starves the hole of fuel, meaning that it turns off. Such cycles may play a key role in how galaxies form, acting like a sort of thermostat. Astronomers now suspect that all galaxies go through phases of being active, perhaps ten per cent of the time. The 'feedback' that results has a strong effect on the subsequent nature of the galaxy.

the condensed idea
Galactic thermostat

34 X-ray background

X-rays are harbingers of extreme physics, and X-ray telescopes flown in space offer views of violent regions, from the neighbourhoods of black holes to the million-degree gas in galaxy clusters. Such objects collectively create a dim X-ray glow across the sky, which is called the X-ray background.

Progress in astronomy often comes from opening up new windows on to the universe. Galileo did this by peering through a telescope; and radio astronomers discovered new phenomena, including black holes, by using radio receivers to pick up the cosmos's signals. At the other end of the electromagnetic spectrum lie X-rays. Decades after radio astronomy's birth, X-ray astronomy was born.

X-rays are generated in extreme cosmic regions that are very hot or pervaded by magnetic fields. These include much of astronomical interest, from galaxy clusters to neutron stars. Yet because they carry so much energy in each photon, X-rays are difficult to collect with a telescope. As we know from their medical use in scans, X-rays travel right through most of the soft tissues in our bodies. If they are fired at a mirror, they don't reflect off but become embedded in it, like a bullet hitting a wall. So reflecting telescopes are impractical for focusing X-rays. Likewise lenses made of glass won't work. The way to control X-rays is to bounce them off a mirror at a narrow grazing angle – they will then be deflected like a ping pong ball, and can be focused. So X-rays can be corralled using series of special curved deflecting mirrors, which are often coated in gold to maximize their reflectivity.

timeline

1895

Röntgen discovers X-rays in lab

Cosmic X-rays X-rays from space are also absorbed by our atmosphere. So astronomers had to wait for the era of satellites to be able to see the X-ray universe. In 1962 an Italian-American astronomer called Riccardo Giacconi and his team launched a detector into space and saw the first X-ray source apart from the Sun, called Scorpius X-1, which is a neutron star. A year later they launched the first imaging X-ray telescope (coincidentally about the same size as Galileo's telescope in 1610). The astronomers made crude observations of sunspots and also imaged the Moon in X-rays.

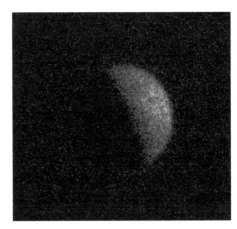

The night sky is brighter in X-rays than the dark side of the moon

The picture of the Moon showed something surprising. The Moon itself was part lit, appearing dark on one side and bright on the other, as you might expect from its phase and from sunlight reflected from its surface. But the sky behind it was not dark – it was also glowing. Catching X-rays is so difficult that these images are built up by individual photons – the background sky showed more photons than the dark side of the Moon, which was obscuring it. Giacconi had discovered the X-ray background.

X-ray background Although both arise at cosmic distances, the X-ray background is different from the microwave background. The former comes mainly from many individual stars and galaxies, which are blended together – in the same way that the Milky Way is made of many stars yet appears as a hazy band to the naked eye. The cosmic microwave background, on the other hand, is due to fossil radiation from the Big Bang that pervades space, and is not associated with particular galaxies.

The quest to work out where these cosmic X-rays were coming from took decades and several further missions. The latest measurements come from NASA's Chandra observatory, which has an eye keen enough to dissect the

1962
Giacconi launches X-ray detector into space

1999
Chandra X-ray Observatory launched

X-ray background. Astronomers have now resolved over 80 per cent of the sources that combine to produce the X-ray background; they suspect the rest is similarly produced but can't yet discern those objects. Forty years after Giacconi's pioneering efforts, over 100,000 X-ray sources have been detected, the most distant of which is 13 billion light years from Earth.

Extreme physics A range of astronomical objects emit X-rays. X-rays are produced in gas heated to millions of degrees. This occurs in regions with high magnetic fields, extreme gravity or in explosions. Some of the largest objects include galaxy clusters – the hot gas that pervades them is spread over a region millions of light years across and can contain enough matter to make hundreds of trillions of stars. Black holes emit X-rays: quasars and active galaxies are very luminous sources and can be traced right across the universe. In fact the presence of a point-like X-ray source in the centre of a galaxy is a giveaway that there is a black hole there.

Astronomers using the Chandra satellite have added X-ray images to multi-wavelength surveys of galaxies, including the Hubble Deep Field and parts of other sky surveys. Using X-ray criteria, they have been able to track the numbers of black holes across the universe over billions of years. These surveys suggest that active galaxies, with accreting black holes, were more common in the past, and that black hole activity has declined since its heyday. This pattern, like the fact that stars were formed more rapidly in the past, may signify that galaxy collisions were prevalent in the young universe.

Some types of stars also glow in X-rays. Exploding stars and supernovae give off energetic emission, as do collapsed stars – stars crushed down by their own gravity as their nuclear burning falters into very dense forms, including neutron stars and white dwarfs. In an extreme case a star will collapse right down to a black hole – X-rays have been detected as close as 90 km from the event horizon of a stellar black hole.

> **❝It seemed at first a new kind of invisible light. It was clearly something new something unrecorded.❞**
> **William Konrad Röntgen**

Young stars, because they are hotter, are stronger in X-rays than our Sun. But the Sun too gives off X-rays in its outer layers, especially its corona, which is super-hot and threaded by strong magnetic fields. X-ray images are useful for looking at turbulence and also the flaring of stars, and how this behaviour changes as stars age. Some of the strongest X-ray sources in our galaxy are close binary systems – pairs of stars – where one or both is a collapsed star. The compact star often sucks gas off the other star, making them very active systems.

WILLIAM RÖNTGEN (1845–1923)

Wilhelm Röntgen was born in Germany's Lower Rhine, moving as a young child to the Netherlands. He studied physics at Utrecht and Zurich, and worked in many universities before his major professorships at the Universities of Würzburg and then Munich. Röntgen's work centred on heat and electromagnetism, but he is most famous for his discovery of X-rays in 1895. While passing electricity through a low-pressure gas he saw that a chemical-coated screen fluoresced even when the experiment was carried out in complete darkness. These new rays passed through many materials, including the flesh of his wife's hand placed in front of a photographic plate. He called the rays X-rays because their origin was unknown. Later, it was shown that they are electromagnetic waves like light except that they have much higher frequency.

the condensed idea
Window on a violent universe

35 Supermassive black holes

Lurking in the centre of most galaxies is a black hole monster. Millions or billions of times more massive than the Sun, and crammed into a region about the size of a solar system, supermassive black holes influence how galaxies grow. The size of the black hole scales with the size of a galaxy's bulge, implying that black holes are fundamental ingredients and may also subject the galaxy to huge blasts of energy if they are activated during galactic collisions.

Since the discovery of quasars and active galactic nuclei in the 1960s, astronomers have known that giant black holes – millions or billions of times more massive than a single star – can exist in the centres of galaxies. In the last decade it has become clear that all galaxies may harbour black holes. In most cases they are dormant; in some circumstances they flare up when material is funnelled on to them, including when we see them as quasars.

There are several ways you can tell whether there is a black hole in the centre of a galaxy. The first is to look at the motions of stars near the galaxy's core. Stars travel on orbits around the centre of mass of a galaxy, in the same way that the planets in our solar system spin about our Sun. Their orbits also follow Kepler's laws, so stars near the centre of a galaxy travel along their elliptical paths faster than those lying further out. The average speeds of stars give away how much mass lies at the centre. The closer in you can measure, the more you can constrain the amount and extent of mass lying within the inner stars' orbits.

timeline

1933	1965
Jansky detects Milky Way centre in radio	Quasars discovered

Astronomers have found that stars in the very centres of most galaxies are moving too fast to be explained by stars, gas and dark matter alone. This is apparent from looking at the Doppler shifts of spectral lines from those innermost stars. The fast stellar motions imply giant black holes at the galaxy's heart – millions or billions of times more massive than the sun, and contained in a region the size of our solar system.

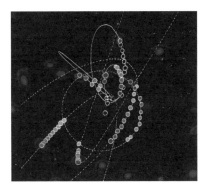

Trajectories of stars near the Milky Way's centre give away the presence of a black hole

Our Galactic Centre The Milky Way has a central black hole. The Galactic Centre lies in the constellation of Sagittarius, near a radio source called Sag A*. Astronomers have tracked dozens of stars near it, and they see clear evidence for a hidden black hole in their movements. Over more than a decade, the stars pursue their orbits, but when they come close to the place where the black hole is thought to lurk, they whip around that point suddenly and are flung back out on elongated paths. Some comets in our solar system take similarly extreme orbits, racing fast past the Sun and slowing down in the icy reaches of the outer solar system. The Galactic Centre stars show that there is something massive, compact and invisible at the centre of the Milky Way whose mass is four million times that of the Sun – a supermassive black hole.

Radio astronomers can similarly measure the speeds of bright sources that inhabit the central regions of galaxies, such as water maser objects, that emit strong radio waves due to excitations of water molecules. In several galaxies the existence of a massive, compact black hole has been deduced using maser velocities that are found to follow Kepler's laws closely.

Bulge–mass relation Before 2000, supermassive black holes were generally thought to be an unusual ingredient in galaxies. Active galaxies clearly had them, and they showed up in some other quiescent galaxies, but they were not considered key. But that quickly changed when astronomers' views of the hearts of galaxies were clarified with new powerful telescopes and instruments that could measure the velocities of stars near their centres. It soon became clear that all galaxies have black holes.

1993
Water masers indicate black hole in galaxy NGC 4258

2000
Black hole–bulge mass correlation discovered

Moreover, the black hole mass is proportional to the mass of the bulge of the galaxy in which it sits. This was the conclusion of a study of hundreds of galaxies in which astronomers measured the spread in the speeds of stars in galactic centres, which indicates central mass, and plotted that on a graph against their bulge mass. It was a nearly one-on-one correlation.

This clear trend was surprising. It held irrespective of the galaxy type, applying to ellipticals as well as the bulges of spirals. It opened up new questions about the relationships between different classes of galaxies, as set out in Hubble's tuning-fork-shaped sequence. Galaxy bulges and ellipticals did not only look alike in terms of their colours and the ages of stars within them; this new correlation suggested that these structures might also have formed in a similar way. It seemed that discs were indeed extra features that might be grown or destroyed, depending on a galaxy's fortunes as it interacted with others.

The scaling is also surprising because proportionally the masses of these black holes are still only a small fraction – less than 1 per cent – of the total mass of the galaxy. So the black hole doesn't influence the galaxy's wider gravitational field as such but is only noticeable in its immediate locale – it is like a black pearl in the galaxy's heart.

Seeds or relics? How might supermassive black holes form? We know that small black holes can arise when massive stars collapse at the end of their lives – when a star ceases burning, it can no longer hold itself up against its own gravity and crunches down to a dense husk. But how would this work on scales millions of times greater? One possibility is that supermassive black holes are the relics of the first stars. The earliest stars to form are likely to have been very large and short-lived, so they would have quickly exhausted their power and collapsed. A cluster of them might be drawn together into a single giant black hole. Alternatively, the black holes

❝[The black hole] teaches us that space can be crumpled like a piece of paper into an infinitesimal dot, that time can be extinguished like a blown-out flame, and that the laws of physics that we regard as 'sacred,' as immutable, are anything but.❞

John Wheeler

in galactic centres may pre-date stars, and might have existed when or soon after the universe was born. It may have been the black holes that seeded galaxies in the first place. Astronomers simply don't know.

The next question is how black holes increase in size. Astronomers think that galaxies grow through mergers – by cannibalizing smaller ones and crashing into large ones. But there are few galaxies with obvious double or multiple black holes within them, even in cases where a merger may have occurred recently. This suggests that central black holes must coalesce quickly – but mathematics and computer simulations imply otherwise. Because black holes are so dense and compact, they behave more like billiard balls than putty. So if they are thrown towards each other in a collision, they should ricochet off rather than stick together. This difference between black hole theory and observation is still a major puzzle.

Feedback Assuming that you can grow black holes smoothly, such that their mass increases alongside that of the bulge in which they sit, how might black holes influence a galaxy? We know that in at least 10 per cent of galaxies the central black holes are active, as we see them as active galactic nuclei. It is plausible that black holes go through phases of activity and dormancy. On average they must switch on, by accreting gas, for around 10 per cent of a galaxy's lifetime. Quasars are clearly affected by the high-energy blast that results – vast outflows of ionized gas, radiation and sometimes radio-emitting particles are generated in the vicinity of a black hole as material pours in. Might all galaxies have been through similar active phases?

Astronomers think so. They suspect that black holes follow cycles of activation in the wake of galaxy collisions. Mergers feed the black hole by bringing in new reservoirs of gas from the other galaxy. The black hole monster then switches on, glowing fiercely in X-rays and blasting out heat and particle outflows. The accumulation of gas also triggers the formation of new stars, so the galaxy goes through a phase of considerable change. Eventually the gas supply is exhausted and the black hole becomes starved and switches off. The galaxy then settles back to its quiescent state – until the next merger. Ultimately, supermassive black holes may be the thermostats that regulate the growth of galaxies.

the condensed idea
A galaxy's black pearl

36 Galaxy evolution

Although Edwin Hubble encapsulated the idea that galaxies change from one type to another by classifying spiral and elliptical galaxies on one diagram, it is still a challenge to figure out how this happens. Astronomers have characterized the different types of galaxies, and mapped the distribution of millions of them across the universe. Now they are running huge computer simulations to try to understand how galaxies form, and how their character depends critically on the basic ingredients of the universe.

The starting point for understanding galaxy evolution is the cosmic microwave background, which is the first snapshot available of the baby universe. The hot and cold spots that fleck its surface map the fluctuations in the density of matter 400,000 years after the Big Bang, which stemmed from tiny irregularities. These seeds then began to grow, through gravity. Clumps of hydrogen gas were pulled together to make the first stars and galaxies.

The next snapshot of the universe that is visible to us is the high-redshift universe. Because of the time it takes light to travel to us, redshifted galaxies are seen as they were billions of years ago. Astronomers can literally see into the past by seeking ever more distant objects. The most distant galaxies and quasars found so far are seen as they were about 13 billion years ago. So we know that galaxies were in place within a billion years of the Big Bang (the universe is 13.7 billion years old). This means that galaxies formed very quickly, well within the billion-year lifetime of an average sun-like star.

timeline

1926	1965
Hubble tuning fork diagram	Cosmic microwave background radiation and quasars identified

❝The reason why the universe is eternal is that it does not live for itself; it gives life to others as it transforms.❞
Lao Tzu

Astronomers have a chicken-and-egg problem when it comes to forming galaxies: did stars form first and bind together to make galaxies? Or did galaxy-sized clumps of gas form first and then fragment into myriads of stars? These scenarios are called 'bottom up' and 'top down' galaxy formation. To distinguish between the two, we need to peer further back in time to find examples of galaxies just forming. This epoch of the universe is difficult to see, however, because it is shrouded in fog – it is referred to as the 'dark age'.

Reionization When the cosmic microwave background photons were released, the universe changed from being electrically charged and opaque (electrons and protons were free to scatter photons) to being neutral and transparent. Atoms formed when the universe cooled enough for electrons and protons to combine, producing a sea of neutral hydrogen with a smattering of light elements thrown in. Yet the universe we see today is almost completely ionized. Intergalactic space is full of charged particles and hydrogen is left over only in galaxies or rare clouds.

What happened to the hydrogen? It was ionized and dissipated when the first stars turned on – a period known as the epoch of reionization. Whether those stars were isolated or already clustered in galaxies should be apparent if we could see the stages in which the ionization happened. But probing the universe's dark age is difficult. First, we know of very few objects with such high redshifts. The most distant galaxies are very faint and red, and searching for them is like looking for a needle in a haystack. Even if a very red object is found, with colours suggestive of being at high redshift, its distance may not easily be ascertained. The characteristic strong lines of hydrogen become redshifted beyond the visible into the infrared, where they are harder to detect. Furthermore, the ultraviolet light that we see

1977	**1992**	**2000**	**2020**
CfA galaxy survey begins	Ripples in cosmic microwave background detected by COBE satellite	Sloan galaxy survey begins	Square Kilometre Array may be operational

redshifted into the visible wavelength range is almost completely absorbed if there is much hydrogen in front of the source. Even so, astronomers think they might have seen a handful of quasars lying on the edge of the reionization epoch, where this absorption is patchy.

In the next decade, astronomers hope to find many more dark age objects. Hydrogen gas also absorbs radio waves at characteristic wavelengths – a key wavelength for a spectral line is 21 cm, which is then redshifted to longer wavelengths according to the object's distance. A new radio telescope that will be built hopes to be able to open up this new low-frequency view of the distant universe. A major international project, the Square Kilometre Array will comprise many small radio antennae scattered over an area of one square kilometre. It will have unprecedented sensitivity, and be powerful enough to map out neutral hydrogen gas structures in the distant universe to locate the first galaxies.

Surveys Hundreds of distant galaxies have been found by their characteristic red colours. Certain types of galaxies stand out more than others – ellipticals and hydrogen-rich galaxies have relatively weak blue and ultraviolet light, which results in a 'step' in their brightness when photographed in a series of adjacent colour filters. Galaxies with very pronounced breaks (due to hydrogen absorption) are called Lyman-break galaxies. At lower redshifts, giant galaxy surveys such as the Sloan Digital Sky Survey have now mapped out much of the proximate universe. We thus have a pretty good view of the recent half of the universe, a sketchier knowledge at high redshifts, a gap in knowledge for the dark age and then a snapshot of the young universe through the cosmic microwave background radiation.

Black hole budget

The role of supermassive black holes in galaxy evolution is a major outstanding puzzle. Astronomers think that most sizeable galaxies harbour black holes, whose mass scales with the size of the galaxy bulge. But black holes are affected by collisions – infalling gas may cause fierce radiation and outflows in the heart of a galaxy; and collisions may kick out black holes rather than slow them enough to coalesce. So the black hole budget still needs to be worked out.

From this information, astronomers are trying to piece together the story of how galaxies and large-scale structures form. Using supercomputers, they are building vast codes to grow galaxies from the first gravitating seeds. The ingredients put in include gas and various types of dark matter, constrained by the initial density fluctuations detected in the cosmic microwave background and the clustering of galaxies seen nearby.

> **'Hierarchy works well in a stable environment.'**
> **Mary Douglas**

Hierarchical models The current favourite model suggests that small galaxies formed first, and collided and merged over time to produce larger galaxies. This is called the hierarchical model. Each galaxy today has a family tree of many smaller galaxies that were swept up within it. Galaxy collisions can be furious and could easily disrupt a galaxy and change its character. Two spirals could crash into each other to leave behind a mess that settles down into an elliptical. That elliptical might then steal a disc from a gas-rich neighbour at a later date. Many types of galaxies can result through simple rules of engagement. On average, though, the sizes of galaxies increase in this model.

Galaxies are not just made of stars and gas – they have dark matter too, spread out over a spherical 'halo'. The nature of the dark matter affects how galaxies collide and cluster. To match the galaxies we see today, the simulations suggest that dark matter shouldn't be too energetic: slow-moving 'cold dark matter' is preferred over fast-moving 'hot' equivalents, which would prevent galaxies from sticking together. Another ingredient is dark energy – a force that works against gravity on large scales. The best results are seen for models that use cold dark matter and that also include a modest degree of dark energy.

the condensed idea
Mighty galaxies from small ones grow

37 Gravitational lensing

Gravitational lensing occurs when massive objects focus light from background sources. Referred to as nature's telescopes, gravitational lenses amplify quasars, galaxies and stars lying behind them, producing multiple images, arcs and occasional rings. Lensing is a powerful tool for astronomy, because it can be used to trace dark material across the universe, including searches for dark matter.

Albert Einstein realized when he developed his theory of general relativity that massive objects distort spacetime. As a consequence, rays of light passing near them travel along curved trajectories rather than straight lines. The resulting bending of light rays mimics the action of a lens and so is called gravitational lensing.

During a full solar eclipse observed in 1919, physicist Arthur Eddington confirmed Einstein's prediction that rays of light bend around masses. Observing a star near the solar limb, Eddington saw that its position was slightly shifted when close to the Sun. If we imagine spacetime as a rubber sheet, with the weighty Sun forming a depression in it, then light rays from a distant star curve around as they pass close by, much as a billiard ball would roll around a dip in a tabletop. When the starlight then arrives at our eye, after its path being diverted by the Sun, it seems to be coming from a slightly different direction.

Einstein presented a theory of gravitational lensing in 1936. A year later astronomer Fritz Zwicky postulated that giant galaxy clusters could act as

timeline

1915	1919
Einstein's relativity theory	Eddington confirms relativity with solar eclipse observation

❝Every body continues in its state of rest or uniform motion in a straight line, except insofar as it doesn't.❞

Arthur Eddington

lenses, their immense gravity distorting galaxies and quasars behind them. But the effect was not discovered until 1979, when a double quasar was spotted – two adjacent quasars with identical spectra.

Multiplying images Multiple images of a single background quasar can be produced by a massive galaxy lying between us and the quasar. The intervening galaxy's mass bends the quasar light as it passes it, channelling the rays into paths around it. In general, such geometries produce odd numbers of images – in the double quasar example above, for instance, a fainter third image should also be visible. The lensed images of the quasar are also amplified. The bending of the rays redirects light forwards from all directions, from the sides of the quasar as well as the front, funnelling it towards us. So images of gravitationally lensed objects can be much brighter than the original.

A cluster's gravity stretches background galaxies into arcs

1936/7	**1979**	**2001**
Einstein and Zwicky predict gravitational lensing	First lensed double quasar confirmed	Microlensing project finds MACHOs towards Magellanic clouds

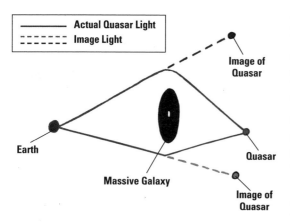

Actual Quasar Light
Image Light

Image of
Quasar

Earth

Quasar

Massive Galaxy

Image of
Quasar

Lenses don't usually line up exactly between us and a distant quasar; such an arrangement results in multiple images as shown in the diagram. But if the background object is exactly aligned behind the lens, then its light becomes smeared out evenly into a circle, called an Einstein ring. If the lens is positioned a little way away, then the ring breaks up into arcs and multiple spots.

A further property of lensed images is that their light rays take slightly different times to travel to us, because they follow different paths. If the background quasar flares up briefly in brightness, then the image with the longest path will experience a slight delay in flaring. Such delays, if the geometry of the lensing system is known, can be used to work out the Hubble constant, the rate of expansion of the universe.

If the background object is a galaxy, which is extended and not a point source such as a quasar, then each part of the galaxy is lensed. The galaxy then appears smeared out and brightened. Because distant galaxies are typically very faint, gravitational lensing can be a powerful tool to reveal the early universe. Galaxies amplified by massive clusters are especially interesting. Such clusters are frequently accompanied by bright arcs, each one signifying a background galaxy smeared out by the cluster's mass. Astronomers can use the geometry of these arcs to ascertain the mass of the cluster, and they can also look at the characteristics of the distant galaxies, which are amplified and stretched.

Weak lensing Multiple images, arcs and rings result when the lens's mass is concentrated and its gravitational effect is high – in the strong lensing regime. But weaker forms of lensing can be detected due to mass that is more spread out and distributed through space. Around the edges of clusters, for example, galaxies tend to be stretched out a little. Because any individual galaxy looks elliptical in shape, it is difficult to say if a particular galaxy is distorted or if it normally looks like that. But on average, patterns can be discerned. Galaxies are slightly elongated due to lensing along the tangent of a circle, or contour, that encloses the mass. So for a circular cluster, galaxies statistically are stretched out so that they tend to form rings around it.

Similarly, a background field of galaxies may be stretched and distorted by more widely distributed matter lying in front of it. The view of the distant universe then appears as if it is seen through an old glass window of uneven thickness, rather than through a clean lens. Astronomers have detected such weak lensing patterns in deep images of the sky by looking for correlations in the orientation of elliptical galaxies. Assuming these correlations are due to lensing, they can work out the distribution of matter in the foreground. In this way they are trying to constrain the distribution of dark matter in space.

Microlensing Another form is microlensing. This happens when small-scale objects pass in front of a background source, or when the lensing mass is quite close to the background object, so that it intercepts its light only partially. Such a technique has been used to look for Jupiter-sized dark matter candidates known as Massive Compact Halo Objects, or MACHOs. In the 1990s, astronomers watched millions of stars towards the Galactic Centre and the Large and Small Magellanic Clouds, tracking their brightness each night for several years. They sought stars that suddenly brightened and then quickly faded in a characteristic way, due to the amplification of a foreground mass. A team observing in Australia found tens of these events, which they attributed to dead stars or rogue gas planets of about Jupiter's mass. Most were seen towards the Galactic Centre, rather than the Magellanic Clouds, suggesting that more of these planetary-sized objects lay within our galaxy than in the Milky Way's outer regions. So the contribution of such MACHOs to the Milky Way's dark matter budget was small. Other dark matter candidates are still sought.

> **❝Something unknown is doing we don't know what.❞**
> **Arthur Eddington**

the condensed idea
Nature's telescope

38 Stellar classifications

The colours of stars tell us about their temperature and chemistry, which is ultimately linked to their mass. In the early twentieth century, astronomers placed the stars in classes according to their hues and spectra, finding patterns that hinted at underlying physics. The classification of stars was an achievement of a notable group of female astronomers who worked at Harvard in the 1920s.

If you look closely, you can see that stars come in many different colours. Just as the Sun is yellow, Betelgeuse is red, Arcturus is also yellow and Vega is blue-white. A cluster of stars in the southern hemisphere was named the Jewel Box by astronomer John Herschel because they glittered like 'a casket of variously coloured precious stones' through his telescope.

What do the colours tell us? Temperature is the main cause of the hues. The hottest stars appear blue, and their surfaces can reach temperatures of 40,000 K; the coolest stars glow red and are only a few thousand kelvins. In between, for progressively cooler atmospheres, a star's colour trends from white to yellow to orange.

This colour sequence mirrors the black body radiation given off by bodies that are ready emitters and absorbers of heat. From molten steel to barbecue coals, the predominant colour with which they glow – the peak frequency of electromagnetic waves emitted – scales with temperature. Stars too emit in a narrow range of frequencies about such a peak, although their temperatures may vastly exceed those of coals.

timeline

1880

Pickering hires Harvard women to map stars

The Harvard computers

The Harvard astronomers who did this were an unusual group for the time. The observatory's leader, Edward Pickering, hired many women to perform the repetitive but skilled tasks that were required to survey hundreds of stars, from making painstaking measurements off photographic plates to performing numerical analyses. Pickering chose women because they were reliable and cheaper to employ than men. Several of these 'Harvard calculators' went on to become well-known astronomers in their own right, including Annie Jump Cannon, who published the OBAFGKM classification scheme in 1901, and Cecilia Payne Gaposchkin, who established that temperature was the underlying reason for the sequence of classes in 1912.

Stellar spectra In the late 19th century, astronomers looked more closely at starlight, splitting it into its rainbow constituents. Just as the spectrum of sunlight shows gaps at particular wavelengths – called Fraunhofer lines – the spectra of stars are striped by dark lines where their light is absorbed by chemical elements in the hot gases that envelop them. The cooler outer layers absorb light produced by the hotter interior.

Hydrogen is the most common element in stars, and so the signature absorption lines of hydrogen are most easily visible in their spectra. The wavelengths that are absorbed reflect the energy levels of the hydrogen atom. These frequencies deliver photons with the right amount of energy to allow the atom's outer electron to jump up from one rung to another. Because the energy levels are spaced out like frets on a guitar, being closer together at high frequencies, the absorption lines that result – corresponding to the differences between the frets – form a characteristic sequence.

For example, an electron in the first energy level may absorb a photon that allows it to jump up to the second level; or the electron may absorb a little more energy and get to the third, or even more and reach the fourth level,

1901	**1906**	**1912**
OBAFGKM stellar classes published	Red giants and red dwarfs identified	Temperature–colour link identified

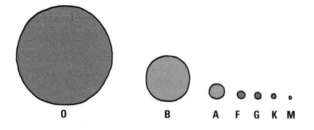

and so on. Each of these steps dictates the frequency of an absorption line. A similar pattern, shifted to slightly higher energies, results for electrons already in the second level; another results for those in the third. For the hydrogen atom, such series of lines are named after famous physicists – the highest-energy one, appearing in the ultraviolet, is called the Lyman series, and the lines in it are known as Lyman-alpha, Lyman-beta, Lyman-gamma and so on. The next series, appearing in the visible part of the spectrum, is the Balmer series, whose primary lines are known (more commonly) as H-alpha, H-beta and so on.

The strength of each of these hydrogen lines depends on the temperature of the gas that absorbs them. So by measuring the relative strengths of the lines, astronomers can estimate its temperature. Other chemical elements in the star's outer layers absorb light, and the strength of their lines may also indicate temperature. Cool stars may have strong absorption lines from heavier elements such as carbon, calcium, sodium and iron. Sometimes they even show signatures of molecules: a common one is titanium dioxide – which is the same chemical that we use in sunblock cream. Heavy elements – which astronomers refer to collectively as 'metals' – tend to make the stars redder.

Magnitudes

In astronomy, the brightness of stars is measured on a logarithmic scale because they span such a great range. The bright star Vega is deemed to have a magnitude of 0; the brightest star Sirius has a magnitude of –1.5; and fainter stars have increasing magnitudes of 1, 2 and so on. The multiplicative factor is about 2.5. If distances are known, then a star's 'absolute magnitude' can be worked out, which is its brightness at a fixed distance, usually 10 parsecs (3.26 light years).

Classification Just as naturalists identified species as a means to understand evolution, astronomers have classified stars according to the characteristics of their light. Initially, stars were categorised by the strengths of various absorption lines, but a more holistic approach was developed at Harvard College Observatory in the US in the late 19th and early 20th century.

The Harvard scheme, still used today, grades stars on the basis of their temperature. From the hottest, approaching 40,000 °K, to cool ones of 2,000 °K, stars are placed into classes named with the letters OBAFGKM. O-stars are hot and blue; M-stars are cool and red. The Sun is a G-type star, with surface temperature around 6,000 °K. This apparently random series of letters arose for historical reasons, from piecing together previous spectral classes that were named either for types of stars or alphabetically. Astronomers often remember it using mnemonics, the best known being 'oh be a fine girl/guy kiss me'. The scheme has since been more finely defined using numbers on a 0–10 scale for sub-classes in between; so a B5 star is halfway between B and A; the Sun is a G2-type star.

> **An attempt to study the evolution of living organisms without reference to cytology would be as futile as an account of stellar evolution which ignored spectroscopy.**
>
> **J.B.S. Haldance**

Although most stars fall into the OBAFGKM categories, some don't. In 1906, the Danish astronomer Ejnar Hertzsprung noticed that the reddest stars had extreme forms: red giants, such as Betelgeuse, are brighter than and have radii hundreds of times that of the Sun; red dwarfs are much smaller and fainter than the Sun. Other types of stars followed, including hot white dwarfs, cool lithium stars, carbon stars and brown dwarfs. Also identified were hot blue stars with emission lines and Wolf-Rayet stars, which are hot stars with strong outflows that show up in broadened absorption lines. The zoo of stellar types suggests that there are laws that might explain stars and their characteristics. Astronomers have had to work out how they evolve – how they change from one type into another as they burn.

the condensed idea
Species of stars

39 Stellar evolution

Stars live for millions to trillions of years. Correlations between their colours and brightness suggest that they follow similar evolutionary paths, dictated by their mass. Their characteristics are due to nuclear fusion reactions going on in their hearts. All the elements around us, including those in our bodies, are the product of stars. We really are made of stardust.

Stars' colours broadly indicate their temperature, such that blue stars are hot and red stars cool. But the typical brightness of stars also varies with colour. Hot blue stars tend to be brighter than cool red ones. Danish astronomer Ejnar Hertzsprung in 1905 and American astronomer Henry Norris Russell in 1913 independently noted similar trends between the brightness and colours of stars. Both astronomers are now recognized in the name of a diagram that plots stars' luminosities against their colours: the Hertzsprung-Russell diagram (or HR diagram for short).

HR diagram On the HR diagram, 90 per cent of stars, including our Sun, lie on a diagonal stripe that runs from bright hot blue stars to fainter cool red ones. This band is known as the main sequence, and stars that fall along it as main sequence stars. Besides the main sequence, other groups of stars are also evident on the HR diagram. These include a branch of red giants – red stars of similar colours but a range of brightness – and a population of white dwarfs – hot but faint stars – as well as a separate branch of Cepheid variable stars, with a range of colours but similar brightness. Such patterns hinted that stars must be born and evolve in consistent ways. But it was not until the 1930s that astronomers understood why stars shine.

timeline
1905/1913

Hertzsprung and Russell
publish trends of star
colours and brightness

❝I ask you to look both ways. For the road to a knowledge of the stars leads through the atom; and important knowledge of the atom has been reached through the stars.❞

Sir Arthur Eddington

Fusion Stars, including the Sun, burn through nuclear fusion – the merging together of light atomic nuclei to form heavier ones, and energy. When pressed together hard enough, hydrogen nuclei can merge to produce helium, giving off a great deal of energy in the process. Gradually, by building up heavier and heavier nuclei through a series of fusion reactions, virtually all the elements that we see around us can be created from scratch in stars.

Fusing together even the lightest nuclei, such as hydrogen, takes enormous temperatures and pressures. For two nuclei to merge, the forces that hold each one together must be overcome. Nuclei are made up of protons and neutrons locked together by the strong nuclear force. This force, which operates only at the tiny scale of the nucleus, is the glue that overpowers the electric repulsion of positively charged protons. Because the strong nuclear force acts only at close range, small nuclei are held together more rigidly than large ones. The net effect is that the energy needed to bind the nucleus together, averaged per nucleon, increases with atomic weight up to the elements nickel and iron, which are very stable, and then drops off again for larger nuclei. Large nuclei are more easily disrupted by a minor knock.

The fusion energy barrier to be overcome is least for hydrogen isotopes, which contain a single proton. The simplest fusion reaction is the combination of hydrogen (one proton) and deuterium (one proton plus one neutron) to form tritium (one proton plus two neutrons) plus a lone neutron. Yet scorching temperatures of 800 million kelvins are needed to ignite even this reaction.

1920
Arthur Eddington proposes
that stars shine by fusion

1939
Hans Bethe works out
physics of hydrogen fusion

1957
Stellar nucleosynthesis
published by B^{2}FH.

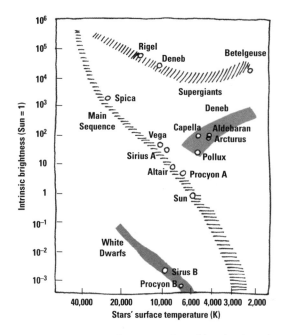

Stardust German physicist Hans Bethe described in 1939 how stars shine by converting hydrogen nuclei (protons) into helium nuclei (two protons and two neutrons). Additional particles (positrons and neutrinos) are involved in the transfer, so that two of the original protons are turned into neutrons in the process. The building of heavier elements then occurs in steps by fusion cookery, in recipes expressed in 1957 in an important scientific paper (known as B^{2}FH) by Geoffrey Burbidge, Margaret Burbidge, William Fowler and Fred Hoyle.

Larger nuclei are constructed through fusing first hydrogen, then helium, then other elements lighter than iron, and in some circumstances, elements heavier than iron. Stars like the Sun shine because they are mostly fusing hydrogen into helium, and this proceeds slowly enough that heavy elements are made in only small quantities. In bigger stars this reaction is sped up by the involvement of the elements carbon, nitrogen and oxygen in further reactions. So more heavy elements are made more quickly. Once helium is present, carbon can be made from it (three helium-4 atoms fuse, via unstable beryllium-8). Once some carbon is made, it can combine with helium to make oxygen, neon and magnesium. These slow transformations take place over most of the life of the star.

The characteristics of a star are further governed by its structure. Stars must balance three forces: their crushing weight, due to their own gravity; the internal pressure of gas and radiation that keeps them puffed up; and the

'We are bits of stellar matter that got cold by accident, bits of a star gone wrong.'
Sir Arthur Eddington

Don't panic

Even if the nuclear reactions in the centre of the Sun switched off today, it would take a million years for the photons produced to reach its surface. So we wouldn't notice what had happened for some time. Even so, there is much historic evidence that the Sun's power remains quite constant.

means by which heat is transported through their gas layers. The first two factors control the structure of the star, which is a series of onion-like layers whose density decreases with distance from the centre. Fusion reactions occur deep in the star's interior, where the pressure is greatest. The heat produced then has to travel through the star to escape at its surface. Heat can be transported in two ways: as radiation, as for sunlight; or through fluid motions of convection, as in boiling water.

Lifetimes The lifetime of a main sequence star is determined by the fusion reaction rates within it and its mass. Reaction rates are very sensitive to the temperature and densities at the star's heart, typically requiring temperatures in excess of 10 million degrees and densities greater than 10,000 grams per cubic centimetre. Massive stars have hotter and denser cores and exhaust their power more quickly than low-mass stars. A Sun-like star lives on the main sequence for about 10 billion years; a star 10 times more massive will be thousands of times as bright, but will only last for 20 million years; a star one tenth of the Sun's mass may be thousands of times more faint, but will last about 1,000 billion years. Because this is longer than the current age of the universe (13.7 billion years), we have yet to see the smallest stars die.

the condensed idea
Star power

40 Stellar births

Stars are born when clumps of gas scrunch down into a tight ball due to gravity. As it collapses, the pressure and temperature of the gas increases until it is high enough to support the star and prevent it from collapsing further. If the mass of the gas ball is high enough, the pressures at the centre become sufficient to ignite fusion reactions and the star turns on.

Most stars form inside giant molecular clouds, reservoirs of dense gas within galaxies. The Milky Way has about 6,000 molecular clouds, accounting for around half of its total gas mass. Nearby examples include the Orion nebula, some 1,300 light years (1.2×10^{16} km) away, and the Rho Ophiuchi cloud complex, 400 light years away. Such regions may be hundreds of light years across and contain enough gas to build millions of suns. They contain a density of gas 100 times that found typically in interstellar space, where one atom per cubic centimetre or less is the norm.

The gas composition of interstellar space is 70 per cent hydrogen, the rest helium, with a smattering of heavier elements. The dense clouds may be cold enough to harbour hydrogen gas molecules (H_2), as well as atoms. Often just a few degrees above absolute zero, molecular clouds include some of the coldest spots in the universe. The Boomerang nebula, for example, has a temperature of just one kelvin above absolute zero, which is lower than the ambient 3 K cosmic microwave background radiation.

Proto-stars Stars are seeded in places within the clouds where the gas density becomes greater than average. Although it is not understood why this happens, it may occur simply due to turbulence or when the cloud is

timeline

> ❝The light which puts out our eyes is darkness to us. Only that day dawns to which we are awake. There is more day to dawn. The Sun is but a morning star.❞
>
> **Henry David Thoreau**

disturbed by a blast from a nearby supernova. Magnetic fields may also play a role in seeding gas clumps.

Once a sizeable clump is formed, gravity kicks in and pulls it together further. As the gas ball concentrates, its pressure rises and it also heats up; gravitational energy is then released, as when a ball accelerates as it rolls downhill. Both forces – heat and pressure – counteract the pull of gravity, and try to halt the sphere's collapse by puffing it up. The critical mass that defines the balance between these two sets of forces is called the Jeans mass, after physicist James Jeans. Clumps that exceed it continue to develop; those that don't will not.

The gravitating region may attract more material from its surroundings, which falls on to it, allowing it to collapse even more. As the clump

Binary stars

Binaries may be identified in various ways – visually, by tracking them through a telescope; spectroscopically, by seeing Doppler shifts in lines indicating that they orbit one another; by eclipsing, where one star dims the other as it moves in front; and astrometrically, where a star is found to wobble slightly, indicating the presence of a companion. William Herschel in the 1780s was one of the first to observe binary pairs of stars, publishing a catalogue of hundreds of them.

1994
Discs identified around forming stars in Orion nebula with Hubble Space Telescope

2009
Herschel space observatory launched

> ❝One must still have chaos in oneself to be able to give birth to a dancing star.❞
> Friedrich Nietzsche

shrinks, it heats up and starts to glow. When its temperature reaches about 2,000 K, it is warm enough to break apart hydrogen molecules and ionize atoms in its host cloud. Offered a new route to release its heat energy, the star is able to collapse still further, and does so until it reaches the point where it is propped up only by its own internal pressure. It is then known as a proto-star.

Proto-stars continue to grow by accreting material. They do so by forming a flat disc, called a circumstellar disc, to funnel the material on to them efficiently. Once the proto-star eats up all the material in its immediate surroundings, it stops growing and contracts again. Eventually it is compact enough to trigger hydrogen fusion in its dense core – it has become a star. For a solar mass star, this process takes 100,000 years. Once it is undergoing fusion, the star has a temperature and colour that place it on the main sequence, where it sits as it evolves according to the patterns dictated by physics.

Forming stars are difficult to observe, because they are faint and buried deep in molecular clouds. Astronomers must look at infrared or longer to catch the dust-obscured glow of proto-stars. Using the Hubble Space Telescope, discs have been spied around massive stars being formed within the Orion nebula; and other observations with 10-metre-class telescopes have

Herschel space observatory

The European Space Agency's Herschel Space Observatory, launched in 2009, is peering at forming stars and distant galaxies at infrared wavelengths. Hosting a large single mirror for a space telescope (3.5m in diameter), it is probing dust- obscured and cold objects that are invisible to other telescopes. Herschel is targeting the first galaxies, clouds of gas and dust where new stars are being born, discs out of which planets may form, and comets. It is named after William Herschel, who identified infrared light in 1800.

similarly revealed discs around individual young stars, confirming that such discs are common. It is an open question, though, whether these discs go on to form planets, such as our own solar system.

Binary stars It is also difficult to explain the formation of twin stars in binary pairs, where both orbit one another about their common centre of mass. About a third of the stars in the Milky Way are in binary pairs – this is too high a rate to have resulted from chance capture of roaming stars, and implies that there must be mechanisms by which double stars form. Clusters of stars might form together if they condense from a single cloud, perhaps simultaneously if it is hit by a shock or disturbance that triggers a mass seeding. Turbulence in the cloud might be a better explanation for lone pairs or multiples formed close together; perhaps others tend to be lost from the system if it is an unstable configuration or through collisions.

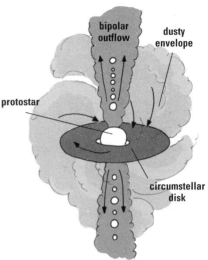

The process by which massive stars are made is another puzzle – they should be much brighter than low-mass proto-stars, so you would expect them to cease collapsing quickly so that they fail to ignite. But on the other hand, they must be easily formed because we see many of them, in particular in places where star formation is vigorous, such as in galaxies following a collision. Perhaps they are efficient in funnelling material on to themselves using a disc and dispelling energy through outflows and jets.

A given molecular cloud might produce stars with a range of different masses. Because each star evolves differently, according to its mass, such a population of stars will look very different over time. For astronomers trying to understand how galaxies form and evolve, the statistics of how stars form affect how the entire galaxy looks.

the condensed idea
The stellar switch-on

41 Stellar deaths

When stars exhaust their nuclear fuel, they burn out. The balance between gravity and pressure that has kept them supported for millions or billions of years is disrupted. As their fusion engine falters, they swell and shed their outer layers; the core crushes down into a compact husk, leaving behind a neutron star, white dwarf or black hole. In some circumstances the star is so destabilized that it explodes as a supernova.

The majority of stars shine for most of their lives by fusing hydrogen into helium nuclei. Whilst they do this, they take on a characteristic colour and brightness that depends on their mass. A Sun-like star glows yellow and sits in the middle of the main sequence, a correlated relationship between brightness and hue adopted by the majority of stars. Stars stay this way for millions of years, brightening and swelling only a little as they age.

Eventually, though, they exhaust their core supply of hydrogen. Counter-intuitively, it is the most massive stars that do so first: hosting much higher pressures and temperatures in their cores, they burn so brightly that the nuclear reactions that sustain them proceed at a rapid rate and they convert their hydrogen in millions of years. Lower-mass stars, in contrast, burn much more slowly and take billions of years to consume their primary fuel.

Last stages When the fusion falters at the centre, the star's helium-rich core contracts and the star heats up as gravitational energy is released. The layers just above the core then start to undergo hydrogen fusion themselves, and dump the helium they generate back into the core. Eventually, the core becomes so dense and hot – reaching 100 million degrees – that it starts to burn its helium, triggering a bright 'helium flash' as fusion switches on again. The helium nuclei combine to produce carbon-12 via one set of

timeline

Tycho's supernova

In early November 1572, a new star appeared in the constellation of Cassiopeia in the northern sky. Spied by Danish court astronomer Tycho Brahe and many others, it was one of the most important sightings in astronomy's history because it showed that the sky changed with time. It also led to improvements in the accuracy by which positions of astronomical objects could be measured. The remnant shell of the supernova was not detected until 1952, and the optical counterpart in the 1960s. In 2004, a companion star to the one that exploded was revealed.

reactions, and also oxygen-16 via another; this is the origin of much of the carbon and oxygen around us. Stars like the Sun can carry on burning helium for about 100 million years.

When helium is exhausted, a similar gear shift may occur, so that the star burns the next element, carbon, in its core, and helium and hydrogen are fused in successive higher shells. But fusing carbon requires even higher temperatures and pressures. So only the largest stars – those exceeding eight solar masses – are capable of entering this phase, during which they become very luminous and bloated. The most massive stars go on to burn oxygen, silicon and sulphur and eventually reach iron.

For lighter stars, less than eight solar masses, the sequence falters when the helium burns out. As the core contracts, episodes of helium and hydrogen burning persist in the layers above, temporarily delivering fuel into the interior of the star. The star goes through a series of bright flashes as fusion switches on and off. While helium is dumped into the centre, the outer layers become distended and are blown off. As the gas within them expands, it cools, and cannot undergo fusion. So the star becomes enshrouded in diffuse cocoons of gas. These bubbles are known as planetary nebulae, because from a distance their circular veils were mistaken for planets. Planetary nebulae don't last for long, however – they dissipate in 20,000 years or so. Some 1,500 are known in our galaxy.

1952	**1987**	**1998/9**
Tycho's supernova remnant discovered	Bright supernova seen in the Magellanic Clouds.	Supernovae used as distance indicators to infer dark energy

Earth

White dwarf

Neutron star

Black hole

Core crush Once these outer layers are shed, the core of the star is left behind. Mostly carbon and oxygen, as everything else is burned or blown off, the hot, dense core quickly fades to a white dwarf. Lacking outward radiation pressure, the material within collapses down into a very compact, dense sphere, equivalent to the Sun's mass being contained within only 1.5 Earth-radii. The resulting density of material is a million times that of water. White dwarfs are supported from becoming black holes only because their atoms cannot be crushed – by quantum electron pressure. They remain very hot, with a surface temperature of 10,000 K. Their heat cannot escape quickly as they have a small surface area, so they survive for billions of years.

More massive stars can compress further. If the remnant exceeds a limit of 1.4 times the Sun's mass (after the outer layers are shed), then the electron pressure is not enough to overcome its gravity and the star collapses to form a neutron star. This limit of 1.4 solar masses is called the Chandrasekhar limit, after the Indian astrophysicist Subrahmanyan Chandrasekhar (1910–95). Neutron stars are confined to a radius of only 10 or so kilometres, squashing the entire mass of the Sun, or several Suns, into a region the length of Manhattan. They are so dense that a sugar-cube-sized block would weigh more than a hundred million tonnes. In the event that the gravity exceeds even this, such as for the largest stars, further compaction ultimately produces a black hole.

Supernovae When very massive stars – tens of times the size of the Sun – die, they may explode as supernovae. After burning hydrogen and helium, massive stars can go through a series of burning stages, working their way up through heavier elements to eventually produce iron. The nucleus of iron is special because it is the most stable nucleus across the periodic table. So when this stage is reached, fusion cannot continue releasing energy by building heavier elements. When this is attempted, energy is absorbed rather than emitted, and the core of the star implodes, passing through the electron-supported white dwarf stage to become a neutron star. When the outer layers fall towards this hard kernel, though, they rebound in a vast explosion of particles (neutrinos) and light.

❝Supernova explosions that are invisible to us because of dust clouds may occur in our galaxy as often as once every 10 years, and neutrino bursts could give us a way to study them.❞
John N. Bahcall, 1987

In a matter of seconds, a supernova gives off many times more energy than the Sun will produce during its lifetime. The supernova is so bright that it briefly outshines the rest of the stars in the galaxy in which it sits, remaining visible for days or weeks before fading from sight.

Supernovae come in two main types, called Type I and Type II. Massive stars cause Type II supernovae. Usually seen in the arms of spiral galaxies, at an average rate of about one every 25–50 years, they show strong hydrogen emission lines due to the shedding of the outer gas layers. The last bright one to go off in our galaxy was spotted by Kepler in 1604. Type I supernovae, however, do not show hydrogen emission lines and are seen in both elliptical and spiral galaxies. They are thought to originate in a different way, in thermonuclear explosions in binary systems that result when a white dwarf is pushed over the 1.4 solar mass Chandrasekhar limit by accreting material from its companion.

Type I supernovae have an important sub-class, known as Type Ia, whose brightness is predictable from following their explosion. By monitoring the way that they brighten and fade, their intrinsic brightness can be inferred, making them useful as distance indicators (see p.54). Because they outshine the rest of their host galaxy, they can be traced across the universe out to high redshifts. Supernovae were used to establish the presence of dark energy.

As iron nuclei are ripped apart in the deaths of massive stars, many neutrons are produced. These can be used to produce other elements heavier than iron, such as lead, gold and uranium. So all these elements on Earth originated in supernovae. Apart from man-made elements, the periodic table is due ultimately from processes that go on in stars.

the condensed idea
Out with a bang

42 Pulsars

Pulsars are spinning neutron stars that send out beams of radio waves. The compact and dense remnants of massive stars, they rotate very fast, completing a revolution in seconds. Their regular signals – originally suggested to be Morse code from aliens – make them accurate clocks, which are important for testing general relativity and detecting gravitational waves.

In 1967, two British radio astronomers picked up a cosmic signal that they couldn't explain. Their radio telescope was crude, yet it broke new scientific ground: it comprised some 120 miles of wire and 2,000 detectors strung across 1,000 wooden posts, like a giant washing line, spread over four acres of a Cambridgeshire field. When it started scanning the sky in July of that year, its pen-plotter spewed out 30 metres of chart each day. PhD student Jocelyn Bell, supervised by physicist Tony Hewish, scoured its graphs to search for quasars that were twinkling due to turbulence in our atmosphere. What she found was something else.

Two months into her observations, Bell spotted a rough patch in the data. It was unlike any other feature and came from one location on the sky. Looking more closely, she saw that it broke up into a regular series of brief radio pulses, one occurring every 1.3 seconds. Bell and Hewish tried to work out where the puzzling signal was coming from. Although its clockwork regularity suggested it might be man-made, they could identify no such emission. It was unlike any known star or quasar.

Little green men? The scientists briefly wondered about a more outlandish possibility: might it be extraterrestrial communication? Although they thought it unlikely to be alien Morse code, Bell recalls

timeline

1967

First pulsar signal picked up

being annoyed that her studies seemed not to be going smoothly: 'Here was I trying to get a PhD out of a new technique, and some silly lot of little green men had to choose my aerial and my frequency to communicate with us'. The astronomers didn't go public, but made more observations.

Bell soon discovered a second pulsating source – dubbed a pulsar – with a period of 1.2 seconds. And by January 1968 she and Hewish had identified four such sources. 'It was very unlikely that two lots of little green men would both choose the same, improbable frequency, and the same time, to try signalling to the same planet Earth,' remarked Bell. More confident that they had detected a new astronomical phenomenon, Bell and Hewish published their discovery in the journal *Nature*.

Neutron stars Astronomers rushed to explain Bell and Hewish's finding. Fellow Cambridge astronomer Fred Hoyle thought it possible that pulses would be given off by a neutron star left behind after a supernova explosion. A few months later, Thomas Gold of Cornell University offered a fuller explanation: if the neutron star was spinning, then a beam of radio waves would sweep past a watching telescope with every rotation, much as a lighthouse beam appears to flash as the lamp rotates.

Nevertheless, it was impressive that a neutron star could rotate once per second. Gold assured them that this was feasible, because neutron stars are so small – only tens of kilometres across. Just after a supernova explosion,

Nobel controversy

Pulsar discoveries have generated Nobel Prizes. Tony Hewish received one, with fellow radio astronomer Martin Ryle, in 1974. Controversially, Jocelyn Bell was not included, despite it being her PhD project that discovered the first pulsar. In 1993, Joe Taylor and Russell Hulse too were awarded a Nobel Prize for their finding of the first pulsar binary system.

1974
Binary pulsar discovered

1982
Millisecond pulsar discovered

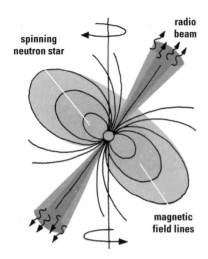

spinning
neutron star

radio
beam

magnetic
field lines

their rapid contraction would cause them to rotate very quickly, in the same way that a spinning ice skater speeds up when they pull in their arms. Neutron stars also possess very strong magnetic fields. It is these that create the twin radio beams, which emanate from opposite ends of the star. As the star spins, the beams sweep out circles on the sky, appearing to flash when pointed at Earth. Gold predicted further that pulsars would gradually slow down, as they lost energy; pulsar spin rates indeed decrease by around a millionth of a second per year.

Gravitational waves The finding of hundreds more pulsars led to further remarkable discoveries. In 1974, American astronomers Joe Taylor and Russell Hulse discovered a binary pulsar: a fast-spinning pulsar that orbited another neutron star every eight hours. This system offered a strong test of Einstein's theory of relativity. Because the two neutron stars are so dense, compact and close together, they have extreme gravitational fields and so offer a new view of curved spacetime. Theorists predicted that as the two neutron stars spiralled in towards one another, the system should lose energy by giving off gravitational waves. By looking for changes in the pulsar's timing and orbit, Hulse and Taylor proved this prediction correct.

Gravitational waves are contortions in the fabric of spacetime that propagate like ripples on a pond. Physicists are building detectors on Earth to detect the squashing of spacetime that is the signature of passing gravity waves, but these observations are extremely difficult to make. Any

Alien map

Although pulsar signals were not sent by extraterrestrials, pulsars feature on the two plaques pinned to the Pioneer spacecraft and on Voyager's Golden Record. In these artefacts, which identify the presence of intelligent life on Earth, for galactic civilizations that might one day find them, the position of the Earth is shown relative to 14 pulsars.

Starquakes

When the crust of a dense neutron star cracks suddenly, it causes a 'starquake', analogous to earthquakes on our planet. These happen as the neutron star compacts down and slows its spin over time, leading to its surface reforming its shape. Because the crust is stiff, it judders. Such quakes have been spotted as sudden drops, or glitches, in the rotation speed of pulsars. Big starquakes can also trigger bursts of gamma rays from pulsars, which can be picked up by satellites including NASA's Fermi observatory.

earthbound shaking, from seismic tremors to ocean wave vibrations, can disrupt the sensitive sensor. Future space missions, using multiple craft placed widely apart and interconnected via lasers, will look for gravitational waves passing through our solar system.

Millisecond pulsars In 1982, another type of extreme pulsar was found: one with a millisecond period (thousandths of a second) was detected by American radio astronomer Don Backer. Rotating 641 times a second, such a rapidly spinning star was remarkable; astronomers think that they arise in binary systems where the neutron star is spun up like a top as it accretes material from its companion. Millisecond pulsars are very accurate clocks: astronomers are trying use them to directly detect gravitational waves as they pass in front of an array of them. Pulsars are certainly useful objects in astronomers' toolboxes.

Pulsars will be one of the main targets of a new-generation radio telescope, the Square Kilometre Array (SKA), a giant array of linked antennae that will start observing in the next decade. By discovering tens of thousands of pulsars, including most of those in the Milky Way, radio astronomers will be able to test general relativity and learn about gravitational waves.

the condensed idea
Cosmic lighthouse

43 Gamma-ray bursts

Gamma-ray bursts are rapid blasts of high energy photons, which occur daily across the sky. First identified by military satellites, most of these bursts mark the last gasps of dying massive stars in distant galaxies. Exceeding the brightness of a normal star whilst located billions of light years away, gamma-ray bursts comprise some of the most energetic phenomena in the universe.

Pulsars and quasars were not the only unusual objects discovered in the 1960s. Unidentified bursts of gamma rays – the most energetic form of electromagnetic radiation – were spotted in 1967 by patrolling US military satellites. Monitoring Soviet compliance with the 1963 Nuclear Test Ban Treaty, which prohibited nuclear tests in the atmosphere, the Vela satellites carried detectors to detect gamma rays given off by nuclear explosions. But the flashes they saw did not look like those from atomic tests. Data on these energetic bursts was declassified in 1973 and published in an academic paper on 'gamma rays of cosmic origin'.

The enormous flashes of gamma rays detected by the satellites appeared from all over the sky. They occurred daily and lasted from fractions of a second to several minutes. These gamma-ray bursts were hundreds of times brighter than a supernova and a billion times brighter the Sun. What was the cause of the energetic flashes?

It took decades to find out where gamma-ray bursts originate. Progress started in 1991 with the launch of the Compton Gamma Ray Observatory satellite, which detected and crudely located thousands of bursts. A plot of

timeline

1967	1991
First gamma ray burst detected by Vela satellite	Compton Gamma Ray Observatory launched

their positions on the sky showed that they were evenly spread (isotropic). They did not come preferentially from the Milky Way's centre or disc, and did not coincide with known extragalactic objects.

This all-sky spread suggests that the gamma rays originated either very near to us or very far away. They don't come from exploding stars across our galaxy, or else they would concentrate in the disc. They could be created locally, but a better guess is that they originate beyond the Milky Way. Yet the fact that they don't cluster near regions of high galaxy density suggests that they come from very far away. This would make them the most energetic phenomena in the universe. The puzzle only deepened.

Gamma-ray bursts come in two different types: long and short duration. Long bursts typically last for tens of seconds; short bursts a fraction of a second. The presence of two distinct classes hints that they are generated by two different processes. Even today astronomers are only just beginning to understand them.

Optical afterglow In 1996, another satellite, BeppoSax, was launched, which made more precise locations possible to derive. As well as detecting gamma rays, the satellite had an X-ray camera bolted on, so astronomers could look for glows at other wavelengths, coincident with the gamma-ray burst. On the ground, they set up an alert system so that when a gamma-ray burst went off, telescopes across the globe could quickly point in that direction to look for any fading counterpart. In 1997, an optical afterglow was spotted, and a very faint galaxy identified as its likely origin. Other afterglow detections soon followed.

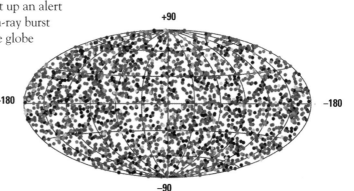

+90

+180

−180

−90

1996
BeppoSax launched

1997
First afterglow spotted

2005
First short-duration-burst afterglow spotted

gamma-ray astronomy

Most gamma-ray astronomy is carried out from space. However, the most energetic gamma-ray photons can be detected by experiments on the ground. As the photons collide with air molecules, they produce showers of particles and flashes of blue light, both of which can be detected. The light – known as Cherenkov light – is most efficiently collected by telescopes. This method has detected gamma rays from the Crab nebula, which hosts a pulsar, and a handful of nearby active galactic nuclei. Although gamma-ray astronomy is difficult, bigger telescopes are being developed that will probe the most violent reaches of space.

With the launch of further satellites, notably Swift and Fermi, astronomers have gathered a range of examples of the counterparts to gamma-ray bursts. Automated telescopes have also been employed, reacting immediately to burst alerts. Clearly these gamma-ray bursts come from very faint distant galaxies, billions of light years away. The association of one burst with a supernova flare implies that long-duration gamma-ray bursts are tied to the death throes of massive stars.

Blast waves Astronomers think that the gamma rays are produced by a blast wave that is generated when the star's core finally collapses to form a black hole. The ensuing explosion sends out a wave travelling at close to

Genius and science have burst the limits of space, and a few observations, explained by just reasoning, have unveiled the mechanism of the universe. Would it not also be glorious for man to burst the limits of time, and, by a few observations, to ascertain the history of this world, and the series of events which preceded the birth of the human race?

Baron Georges Cuvier

the speed of light, which passes through gas remaining around the star, generating gamma rays just ahead of the shock front. Other forms of electromagnetic radiation are also made in the blast wave, producing the afterglow, which may linger for days or weeks.

Short bursts have posed more of a problem for identification because any afterglow might have disappeared before a telescope can be slewed to look in its direction. Since 2005, though, a handful of associations with short bursts have been spotted. However, these were found in regions lacking active star formation, including in elliptical galaxies, which suggests that short bursts are physically different and not simply due to massive star deaths. Although their origin is still unclear, it's thought they might arise when neutron stars merge, or in other energetic systems. Gamma-ray bursts are typically one-off, catastrophic events – only a handful ever repeat.

> **'Even one well-made observation will be enough in many cases, just as one well-constructed experiment often suffices for the establishment of a law.'**
> Émile Durkheim

Particle beam Gamma-ray bursts give off more energy than any other known astrophysical object. They shine temporarily like a bright star even though they lie billions of light years away. Astronomers find it difficult to explain how so much energy can be given off so quickly. One possibility is that in some cases the energy is not emitted equally in all directions but, like pulsars, electromagnetic waves are given off primarily in a narrow beam. When that beam is directed at us, we see a high-energy flash. The gamma rays might be amplified too by relativistic effects if they arise from fast-moving particles spiralling in magnetic fields, possibly in small-scale versions of the particle jets that emanate from radio galaxies. So the mode of generation of gamma-ray bursts is still under investigation.

Given that gamma-ray bursts occur billions of light years away, yet appear as bright as a nearby star, we are lucky that they are so rare. If one went off in our neighbourhood, it might fry the Earth.

the condensed idea
Giant flashes

44 Variability

Astronomers are opening up new views of the universe by looking at how objects vary with time. Most stars shine constantly. However, others – variable stars – do undergo physical changes that cause them to vary in brightness. The way in which their light fluctuates can reveal much about the star. The cosmos is a place of change.

Although comets and supernovae have surprised people across the centuries by being celestial visitors, the night sky has generally been considered unchanging. This picture of constancy was altered in 1638 by Johannes Holwarda's discovery of the pulsations of the star Mira, which brightens and fades on an 11-month cycle. By the late 18th century, a handful of variable stars were known, including Algol. The number rapidly increased after the mid 19th century, as photography made monitoring of large numbers of stars routine. Today, over 50,000 variable stars are recognized; the majority lie in our galaxy, but many have been detected in other galaxies.

Pulsations Variable stars come in many guises. Monitoring of the light output of a star reveals how its brightness rises and falls – its light curve. The cycle may be periodic, irregular or somewhere in between. The star's spectrum also tells us its type, its temperature and mass, and whether it is a binary or not. Spectral changes may accompany the fluctuations in the star's light. Spectral lines can show Doppler shifts that indicate the expansion or contraction of shells of gas, or the presence of magnetic fields. Once all the evidence is amassed, the reasons for a star's variability can then be deduced.

Around two thirds of variable stars pulsate – they swell and contract in regular cycles. Such behaviour arises due to interrelated instabilities in the

> **"Scientific development depends in part on a process of non-incremental or revolutionary change."**
>
> Thomas S. Kuhn

star, which cause it to oscillate. One mode, pointed out by Arthur Eddington in the 1930s, is driven by changes in the degree of ionization of the star's outer layers, which are coupled to its temperature. As the outer layers swell, they cool and may become more transparent. It then becomes easier for the star to radiate more energy, so it contracts. This heats the gas again, which makes the star again start to swell. This cycle repeats over and over.

Cepheid variables Such a pattern explains the pulsation of Cepheid variables – an important type of variable star used as a distance indicator. Cepheid cycles are driven in particular by changes in the ionization of helium. Doubly ionized helium is more opaque than singly ionized helium; so oscillations in transparency and temperature result. The period of these cycles is closely related to the luminosity of the star.

Quasar variability

Variability is not confined to stars. Many quasars are variable. Their variability, together with their uniform brightness across the electromagnetic spectrum, has been used as a means of finding them. Quasar variability may be due to changes in the amount of material being accreted on to their central super-massive black hole, or due to a hotspot on their accretion disc whose brightness changes. The fastest variability timescale seen in quasars tells us about the size of the region producing that light. For instance, if quasars vary on timescales of days, then a light day can be estimated as the smallest size for that structure, such that light can communicate coherently across that distance.

1908
Cepheid period-luminosity relation worked out

1924
Cepheids used to measure distance to Andromeda nebula

2014
Large Synoptic Survey Telescope opens

Size and colour

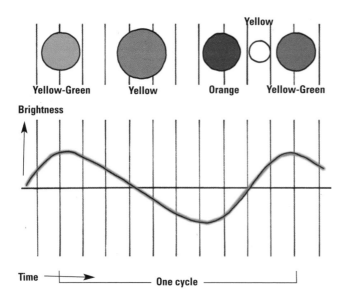

Yellow-Green   Yellow   Orange   Yellow   Yellow-Green

Brightness

Time ⟶ — One cycle —

Cepheids are very luminous massive stars – they are typically 5–20 times as massive as the sun and up to 30,000 times as luminous. They may vary on timescales of days to months, during which their radii change by almost a third. Their brightness and predictable variability means that they can be seen out to distances of 100 million light years. Thus they can be traced in nearby galaxies, and their luminosity can be ascertained, making them good distance indicators.

Cepheid variables were discovered in 1784, and are named for the prototypical star, Delta Cephei. A better-known example is Polaris, the north pole star. The period–luminosity relationship was discovered in 1908 by Harvard astronomer Henrietta Swan Leavitt, based on observations of Cepheids in the Magellanic Clouds. Cepheids were a critical part of the jigsaw puzzle in establishing the size of the Milky Way, and distances to galaxies beyond our own. In 1924 Edwin Hubble used them to work out the distance to the Andromeda galaxy, clearly showing that it lay beyond the Milky Way. Cepheids have also played a key role in measurements of the expansion rate of the universe, via Hubble's Law.

> In the past century [19th], there were more changes than in the previous thousand years. The new century [20th] will see changes that will dwarf those of the last.

H.G. Wells

Sky movies

In future, time-variable astronomy will become routine. The sky will be monitored like a movie, rather than as a series of snapshots. The next generation of telescopes – both optical and radio – are being designed to provide continuous surveying of the sky, making searches for new types of variable objects – and hopefully many surprises – possible. One such telescope is the Large Synoptic Survey Telescope, due to open in Chile in 2014. With an 8.4 m diameter mirror and a wide field of view, it will survey the entire sky twice a week, taking 800 images each night. Each patch of sky will be visited 1,000 times in 10 years. Several billion stars and billions of galaxies will be imaged. As well as variable stars and quasars, numerous supernovae should be picked up, allowing for tests of dark energy.

Cepheids are one type of intrinsically variable star. Such stars physically deform to produce variability. In the case of Cepheids, it is through pulsation; other stars may appear variable due to eruptions or flares on their surfaces. Yet others vary as a result of extreme processes that lead to explosions, such as cataclysmic variable stars, novae and supernovae. Alternatively, extrinsically variable stars may exhibit eclipses from an orbiting companion, or may have singular marks on their surfaces, including giant sunspots, that cause variability as the star rotates. Most classes of variable star are named for their prototype, such as RR Lyrae stars, which are like Cepheids but fainter, and Mira variables, which pulsate due to hydrogen ionization changes rather than helium.

the condensed idea
All-sky movie

45 The Sun

Our nearest star, the Sun still holds mysteries. While it has revealed much about the process of nuclear fusion and the structure of stars, its magnetic weather can be unpredictable. Following an 11-year cycle of activity, the Sun is subject to erratic flares and solar wind surges. These can paint beautiful aurorae on Earth, disrupt our electronic communications systems, and affect our climate.

The ancient Greeks recognized that the Sun was a giant ball of fire lying far from Earth. But it wasn't until the 16th and 17th centuries that it was demonstrated that the Earth orbits the Sun, and not vice versa. The advent of the telescope in the 17th century revealed sunspots, dark blotches moving across the face of the Sun. Galileo Galilei observed them and realized that they were storms on its surface, and not intervening clouds. In the 19th century, the chemical composition of the Sun was established by identifying dark absorption lines in the solar spectrum – Fraunhofer lines. But knowledge of what powered the Sun – nuclear fusion – wasn't obtained until the twentieth century, when atomic physics was developed.

The Sun contains most of the mass of the solar system (99.9 per cent) in a ball whose diameter is about 100 times that of the Earth. It is about 150 million kilometres away, and light from it takes eight seconds to reach us. About three quarters of the Sun's mass is in hydrogen, the remainder being helium with heavier elements such as oxygen, carbon, neon and iron thrown in. It burns due to nuclear fusion of hydrogen into helium in its core. With a surface temperature of 5,800 K, the Sun is a yellow star of G2 classification, of average brightness for a main sequence star, and about halfway through its 10 billion year life.

timeline

1610	1890	1920
Galileo publishes telescope observations	Joseph Lockyer discovers helium in the solar spectrum	Arthur Eddington proposes that fusion powers the Sun

> **The Earth in its rapid motion round the sun possesses a degree of living force so vast that, if turned into the equivalent of heat, its temperature would be rendered at least one thousand times greater than that of red-hot iron, and the globe on which we tread would in all probability be rendered equal in brightness to the sun itself.**
> James Prescott Joule

Solar structure The Sun has an onion-like structure. At its heart, comprising the innermost quarter of its radius, is the hot, dense core. Fusion occurs here, releasing energy equivalent to consuming four million metric tons of gas per second; or exploding tens of billions of megatons of TNT per second. Temperatures in the core reach a scorching 14 million kelvins. The next layer is the radiative zone, which lies between 0.25 and 0.7 solar radii. Energy from the core travels through this region as electromagnetic radiation – photons. Temperatures here decrease outwards, from 7 million to 2 million kelvins.

Above the radiative zone lies the convection zone, accounting for the outer 30 per cent of the Sun's radius, up to the surface. Heat rising from below causes gas here to bubble up to the surface and plunge back down again, circulating like water brought to the boil in a pan. Heat is rapidly lost from the region, so the surface temperature drops to 5,800 K. A thin surface layer – the photosphere – coats the Sun; it is only a few hundred kilometres thick.

Tenuous gas above the surface forms the solar atmosphere, which can be seen during a full eclipse of the Sun by the Moon. It comprises five regions: a cool layer 500 km thick known as the temperature minimum region; the chromosphere, a hot ionized region 2,000 km thick; a 200 km thick transition region; the extensive corona, which stretches far from the Sun and generates the solar wind and is very hot, reaching millions of degrees;

1957	**1959–68**	**1973**	**2004**
Burbidge *et al.* work out theory of stellar nucleosynthesis	NASA Pioneer probes observe solar wind and magnetic field	Skylab launched; observes solar corona	Genesis captures solar wind particles

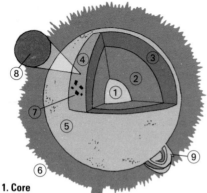

1. Core
2. Radiative zone
3. Convective zone
4. Photosphere
5. Chromosphere
6. Corona
7. Sunspot
8. Granules
9. Prominence

and the heliosphere, a bubble filled with solar wind stretching to the edge of the solar system. In 2004, the Voyager spacecraft passed through the edge of this bubble, travelling through a shock front called the heliopause.

Space weather The Sun possesses a strong magnetic field. It reverses direction every 11 years – marking the solar cycle – and also undergoes continual changes. More frequent sunspots, flares and bursts of solar wind arise when the Sun's magnetic field is particularly active. Such outbursts can send clouds of particles whizzing through the solar system. When they reach Earth, they are funnelled by the Earth's own magnetic field on to high-latitude regions, where they glow as delicate aurorae, or the Northern and Southern Lights. Powerful eruptions of particles can be destructive, knocking out telecommunications and power grids, as happened in Quebec, Canada, in 1989.

Sunspots are vortexes of strong magnetic field that arise on the surface of the Sun. Reaching thousands of kilometres across, they appear dark because they are colder than the surrounding boiling gas. Sunspot numbers increase when magnetic activity peaks, fluctuating every 11 years or so. Unusual solar cycles can affect Earth's climate: the Little Ice Age that froze Europe in the 17th century coincided with the solar cycle stopping for several decades; few sunspots were observed at all during this period. In the few years running up to 2010, the Sun has been in a quiet phase – its brightness has dropped slightly and its magnetic field, number of sunspots and solar wind strength are all lower than average.

Genesis

Because only the outer layers of the Sun absorb light, the chemistry of the interior is barely known. A space mission called Genesis collected particles from the solar wind to measure its composition. In 2004 it plunged back to Earth, carrying the samples with it. Although its parachute failed and it crashed into the Nevada desert, astronomers have managed to piece together the fragments of its detectors to analyse the particles from the Sun.

Puzzles The Sun is a good laboratory for stellar physics. Although we know a lot about how it works, there are still many puzzles. One that was only recently solved is the mystery of the missing solar neutrinos. Nuclear fusion of hydrogen to helium produces particles called neutrinos as a by-product. These should be given off in vast numbers by the Sun – but physicists could only see less than half the quantity they expected. Where were the rest? Neutrinos are difficult to detect because they hardly interact with matter. In 2001 the Sudbury Neutrino Observatory in Canada gave the answer: the reason for the shortfall was that the neutrinos were morphing into other sorts of neutrinos during their journey from the Sun. Physicists detected these other versions (tau and muon neutrinos), demonstrating that the neutrinos 'oscillated' between these types and that the particles had a measurable, if tiny, mass (rather than being massless as was previously thought). The solar neutrino accounting problem was solved.

> **And teach me how To name the bigger light, and how the less, That burn by day and night ...**
>
> **William Shakespeare**

A second solar puzzle is still unexplained: the heating mechanism for the Sun's million-degree corona. The photosphere is only 5,800 K, so the corona isn't heated by radiation from the Sun's surface. The best guess so far is that magnetic energy pervades the corona's plasma. It is created when magnetic fields lines snap, crackle and pop, through flares and magnetic waves that traverse the gas.

The Sun's fate The Sun is about 4.5 gigayears old, and about halfway through its life cycle. In another five billion years it will exhaust the hydrogen fuel in its core and swell to become a red giant. Its bloated outer layers will extend beyond Earth's orbit, 250 times the present radius of the Sun. Although the planets may be loosened to drift into more distant orbits as the Sun loses mass, the Earth may not be spared. Our oceans and water will be boiled away and the atmosphere lost. Even now, the Sun is increasing in luminosity by about 10 per cent every billion years, so terrestrial life may become extinct within a billion years or so. The Sun will end its days as a white dwarf, having shed its gas layers to appear temporarily as a planetary nebula. Only the core will be left.

the condensed idea
Our nearest star

46 Exoplanets

Hundreds of planets are now known around stars other than the Sun. The majority found so far, revealed by the spectroscopic wobble they give to their parent star, are gas giants like Jupiter. But space missions are seeking smaller rocky planets that may be habitable analogies of Earth.

The discovery of planets around stars other than the Sun – exoplanets – has been a holy grail of astronomy. Given that there are so many stars in the Milky Way, it seems unlikely that our solar system is the only one. Yet detecting dim bodies orbiting bright distant stars has proved difficult, and exoplanets were not spotted until the 1990s, when instruments on telescopes improved sufficiently to reveal them. A flurry of detections followed – over 400 exoplanets are now known.

Barring a handful of planets located around pulsars by radio astronomy techniques, the bulk were picked up by their signatures in the spectra of stars. In 1995 Michel Mayor and Didier Queloz of the University of Geneva made the first such detection when they perfected the method of looking for slight shifts in the wavelengths of starlight, due to the tug of a planet upon the star.

Finding planets   Because two massive bodies are orbiting one another about their joint centre of mass – a point that lies closer to the more massive body, rather than midway between them or centred on either – the planet's presence causes the star to inscribe a small circle as its companion moves around it. This wobble can be picked up as a Doppler shift in the star's light: as the star moves away from us its light is shifted to redder

timeline

1609	**1687**	**1781**
Kepler publishes theory of orbits as ellipses	Newton explains Kepler's laws with gravity	William Herschel discovers Uranus

> **❝A time will come when men will stretch out their eyes. They should see planets like our Earth.❞**
>
> **Sir Christopher Wren**

wavelengths; as it moves towards us it appears a little bluer. Even though we cannot see the planet itself, we can detect its presence as its mass causes the star to dance back and forth (see Doppler effect p.32).

The majority of exoplanets identified so far have been found using this Doppler method. In theory we could look for the star's wobble directly as a small change in its position; but such a fine measurement is extremely difficult to make because stars are so far away. Another method is to look for regular dimming of a star due to transits of a planet in front of it. A planet like the Earth would block a tiny fraction (about 100 parts per million) of the star's light for several hours at a time. For a robust detection, this drop must repeat reliably, on a cycle that may last days or months to years. Once its orbital period is measured in this way the planet's mass can be calculated using Kepler's Third Law. A handful of planets have been found this way so far.

Kepler mission

Launched in 2009, NASA's Kepler spacecraft is designed to find Earth-like planets. Its 0.9 m diameter telescope continually watches a large swath of sky (105 square degrees) that includes 100,000 stars. Should an Earth-sized planet pass in front of any star, it will be picked up as a dip in its brightness. Over three and a half years, the mission hopes to detect hundreds of such planets, or place a limit on their numbers if few are found.

1843–6	**1930**	**1992**	**1995**	**2009**
Neptune predicted and found by Adams and Le Verrier	Clyde Tombaugh discovers Pluto	First extrasolar planet discovered around a pulsar	First exoplanet found by Doppler method	Kepler mission launched

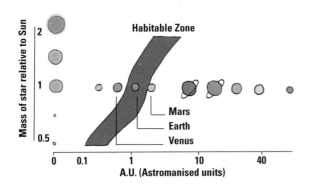

Different detection methods tend to pick out different types of planets. The Doppler method is most sensitive to very large planets, like Jupiter, orbiting very close to their star, where they exert the strongest pull. The transit method can track more distant and smaller planets, including Earth-like ones, but requires very sensitive measurement of the star's light, over lengthy periods. This is best done from space, above the Earth's turbulent atmosphere. The transit method is being employed by several missions, including NASA's Kepler spacecraft, which was launched in 2009.

Hot Jupiters Of the hundreds of planets detected so far, most are gas giants lying very close to their parent star. They have masses similar to that of Jupiter – and almost all are more than 10 times the Earth's mass – but move in very tight orbits that lie much closer to their star than Mercury is to our Sun. These 'hot Jupiters' typically orbit their stars in just a few days and their atmospheres become hot due to their proximity. One planet has been shown to have a hotter day side, reaching 1,200 K when facing the star, and a cooler night side, dropping to around 970 K. Astronomers have detected water, sodium, methane and carbon dioxide in the spectra of exoplanet atmospheres.

Exoplanets are defined as being orbiting bodies of too small a mass to undergo deuterium fusion – they are not big enough to ignite and become stars. In practice the maximum size is around 13 times that of Jupiter. Inactive balls of gas larger than this fusion limit are called brown dwarfs. There is no lower mass limit, other than the typical scale of planets in our solar system. Exoplanets may be gas giants like Jupiter and Saturn or rocky like Earth and Mars.

Found around about 1 per cent of the main sequence stars investigated so far, exoplanets are common. Even if this statistic is an underestimate, as seems likely given the observational bias for hot Jupiters, it implies that there must be billions of planets in the Milky Way, which contains

100 billion stars. Some stars are more likely to host planets than others. Stars similar to our Sun (spectral classifications F, G or K) are most likely to do so; dwarf stars (class M) and luminous blue stars (class O) are less likely. Stars whose spectra show they contain relatively more heavy elements, such as iron, are more likely to have planets – and massive ones at that.

> **The only truly alien planet is Earth.**
>
> J. G. Ballard

Many of the orbits of exoplanets detected so far are extreme. The most rapidly orbiting ones, circling their star in less than 20 days, tend to follow near-circular tracks, similar to those seen in our solar system. Those taking longer tend to follow elliptical and sometimes highly elongated orbits. That these stretched orbits persist, and do not settle into circular ones, is difficult to explain. Nevertheless, it is remarkable that the same physics applies to these distant planets as to our own solar system.

Habitable zone To map out the planetary systems of other stars, astronomers are keen to find planets of lower mass that lie further away from their host star than the hot Jupiters. Earth-like planets are especially sought – rocky exoplanets with similar masses and locations relative to their star as Earth is to the Sun. Around each star is a 'habitable zone', where a planet lying at such a distance from its star would be of the right temperature to host liquid water, and thus the possibility of life. If a planet lies any closer, then water on its surface would boil off; any further away from the star and water would be frozen. The key distance depends on the brightness of the star – habitable planets lie further away from bright stars, nearer for faint ones.

Astronomers have indeed learned a lot in the last twenty years about planets. The ultimate holy grail still awaits them: finding an Earth analogue planet around a distant star. But as technology and observational accuracy progresses, it is only a matter of time before entire exoplanetary systems will be mapped.

the condensed idea
Other worlds

47 Formation of the solar system

The Sun formed from a giant cloud of gas 4.5 billion years ago. Just as other stars condense out of molecular clouds, so the Sun grew gravitationally out of a sea of hydrogen, helium and traces of other elements. The planets formed from the debris left over. Accretion and collisions dictated their sizes and positions in a game of cosmic billiards.

When the heliocentric model gained acceptance in the 18th century, questions about the origin of the solar system arose. The idea that the Sun and planets formed from a giant cloud of gas – the nebular hypothesis – was put forward by Emanuel Swedenborg in 1734, and developed later that century by Immanuel Kant and Pierre-Simon Laplace. While broadly true, the picture has developed much since then. Just as other stars form from molecular clouds, such as the Orion nebula, the Sun must have condensed from a cloud rich in hydrogen, helium and traces of other elements.

The pre-solar cloud would have been many light years across and contained enough gas to potentially make thousands of Suns. The Sun may not have been alone in this cloud – meteorites containing quantities of a heavy isotope of iron (Fe-60) suggest that the nebula was polluted by ejecta from a nearby supernova. Thus the Sun might have grown up amongst other massive stars, which would have been short-lived and exploded before the solar system came into being.

The Sun grew gradually from an overdense region of the cloud due to gravity. In 100,000 years it became a proto-star – a hot, dense ball of gas not yet undergoing fusion. It was surrounded by a circumstellar disc of gas and

timeline

1704

First use of phrase 'solar system'

Comet crash

From 16–22 July 1994, Comet P/Shoemaker-Levy 9 crashed into Jupiter's atmosphere. This was the first collision of two solar system bodies seen, and was watched from most observatories on Earth and in space. By the time the comet approached Jupiter, its nucleus was torn into at least 21 fragments up to two kilometres in size. Astronomers watched as the pieces hit the atmosphere one by one, triggering plumes and fireballs.

dust that stretched to several hundred times the current Earth's radius. After around 50 million years, the Sun's fusion engine switched on and it became a main sequence star.

Growing planets The planets formed from debris collected in the disc. Grains coalesced and clumped up to make objects kilometres in size, and these subsequently collided and stuck together. The planetary embryos grew larger and larger. At the same time, regions of the disc became cleared of material near where the planets were forming.

The inner regions of the forming solar system were hot, so volatile compounds like water could not condense there. Rocky, metal-rich planets formed based on chemicals with high melting points – iron, nickel and aluminium compounds and silicates, the mineral bases of the igneous rocks we see on Earth today. The terrestrial planets – Mercury, Venus, Earth and Mars – grew steadily as smaller bodies merged. It is thought that the inner planets formed further from the Sun than they lie today – their orbits

'The Sun grew gradually from an overdense region of the cloud due to gravity' – the overdensity will have collapsed under its own weight, and that grew due to gravity attracting more gas.'

Tycho Brahe

1734
Swedenborg proposes nebular hypothesis

1994
Comet crashes into Jupiter

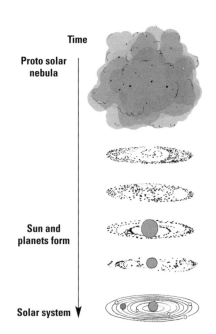

Time

Proto solar nebula

Sun and planets form

Solar system

contracted as the planets were slowed by drag from gas remaining in the disc, which eventually dissipated.

Giant gas planets – Jupiter, Saturn, Uranus, Neptune – formed further out beyond the 'ice line', where volatile compounds remain frozen. These planets were large enough to mop up hydrogen and helium atmospheres; the four of them together comprise 99 per cent of the mass orbiting the Sun. After 10 million years, the young Sun had blown away all the extraneous gas in the disc so that the planets remained and stopped growing.

It was originally thought that the planets were formed largely in the positions we see them today. But in the twentieth century, astronomers realized that this was not so. They developed new theories that suggested that the planets had in fact moved around a lot due to collisions, in a cosmic game of billiards.

Giant impacts When the inner planets were nearly complete, the region was still littered with hundreds of Moon-sized planetary embryos. These collided with the established planets in giant impacts. We know that such events occurred: Earth gained its Moon in one collision; Mercury lost much of its outer shell in another. The most likely reason for this number of collisions was that the planets' orbits were more elongated than they are now, so they crossed paths with smaller objects frequently. Since then the orbits have regularized, becoming near-circular perhaps through successive collisions or drag from debris.

The rubble of the asteroid belt, between Mars and Jupiter, may be the remnants of a planet that was shattered by many collisions. The region was particularly prone to disruption due to the gravitational influence of Jupiter, the largest planet in the solar system. When Jupiter's orbit shifted, it caused widespread disruption. Gravitational 'resonances' stirred up the region just inside its orbit. The resulting collisions shattered the planet that was there, leaving behind scattered asteroids. Some icy asteroids from this belt might have been flung into Earth's orbit, delivering water to our young planet. Water may also have come from comets.

Meteorites

Meteorites are made up of cosmic debris, including material left over from the early solar system and planetary shards. There are three main types. Iron-rich meteorites come from the cores of shattered asteroids; stony meteorites are mostly made of silicates; and stony-iron meteorites are a mixture of both.

The minerals in these dark rocks contain isotopes, whose ratios can be used as cosmic clocks to measure when they were formed by their rates of radioactive decay. By piecing together these timings, the way in which the building blocks of the solar system were distributed and put together can be ascertained.

Jupiter and the other outer planets moved around a great deal in the late stages of their formation. The disc would have been too cold and diffuse at the radii of the outermost planets for sizeable objects to form. So Uranus and Neptune, and Kuiper Belt objects including Pluto and comets, must have formed closer to the Sun and been flung out by gravitational interactions. Neptune may even have formed within Uranus's orbit, and been shuffled out beyond it. A possible reason is an orbital dance that began between Jupiter and Saturn 500 million years after the solar system's birth. For a period, Jupiter's orbit was twice as fast as Saturn's, causing resonant tidal ripples that rang throughout the solar system. Neptune was pushed out, and small icy bodies scattered into the Kuiper Belt.

> **It took less than an hour to make the atoms, a few hundred million years to make the stars and planets, but five billion years to make man!**
>
> George Gamow

Late bombardment During the period when the outer planets were shuffling about, lots of asteroids were flung into the inner solar system. The orbits of the terrestrial planets were by now relatively settled, the major collisions having ended. A period of 'late heavy bombardment' resulted, during which many impact craters were formed on the Moon and the surfaces of other planets were scarred. It was after the bombardment ceased, 3.7 billion years ago, that the first signs of life emerged on Earth.

the condensed idea
Cosmic billiards

48 Moons

Apart from Mercury and Venus, all other planets in the solar system have one or more moons. Many poets have mused on the beauty of our own Moon, but imagine how dramatic the scene would be if there were more than fifty orbs in our skies, as Saturn and Jupiter each has. Moons can be formed in three ways: *in situ*, having grown from a disc of gas and rubble around a planet; through capture of a passing asteroid; or chipped off the planet through a violent impact with another body. Such a collision may be responsible for our Moon.

The giant outer planets are so vast that they retain orbiting debris. Jupiter, Uranus and Neptune all have rings, but Saturn's are by far the largest and have been puzzled over since the 17th century, when Galileo peered at them through his telescope. Thousands of rings circle Saturn, reaching out nearly 300,000 km from the planet, and all lying in a thin plane just a kilometre thick. The rings are made up of billions of small lumps of ice, ranging from the size of a sugar cube to that of a house.

Saturn has more than 50 moons, and each is unique. Titan, the largest, discovered in 1655 by Dutch astronomer Christiaan Huygens, hosts a thick and orange-tinted atmosphere that is composed mainly of nitrogen. Iapetus appears white on one side and dark on the other, as ice coats its front as it moves through the ring material; Mimas has an enormous crater on one side from a past collision; and Enceladus is active beneath its surface, shooting columns of water vapour from its ice volcano. Tens of smaller moons have been detected, many having carved out gaps in the ring system as they formed by accreting icy shards.

The inner planets are too small to have grown moons from rings of debris – they have captured them. Mars's moons Deimos and Phobos are thought to

timeline

be acquired asteroids. In the Earth's case, the creation of the Moon was more violent. In the early solar system, when many sizeable bodies were crashing around as planetary embryos formed, a passing asteroid is thought to have directly hit the Earth. The Moon is the result of that clash.

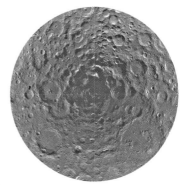

Giant impact hypothesis Although the question of the origin of the Moon has long been of interest, it received renewed attention in the 1970s during the Apollo programme. Astronauts brought back rocks and geological information, and planted detectors on the Moon's surface to pick up seismic signals and reflect laser light to establish the Moon's accurate distance from Earth. They found that the Moon is retreating from Earth at a rate of 38 mm per year, and that it has a relatively small partially molten core. The composition of the Moon's crust is very similar to igneous rocks on Earth.

For a long time scientists had thought the Moon was formed at the same time as the Earth, when a droplet of molten magma was spun off. But the small size of the Moon's core – 20 per cent of the radius of the satellite, compared with 50 per cent for Earth – suggested that a different explanation was needed. If the Moon had formed at the same time as the Earth, its core should be much larger. In 1975, William Hartmann and Donald Davis worked up an alternative hypothesis, which suggested that another body was involved in creating the Moon in a near-catastrophic impact.

> **Here men from the planet Earth first set foot upon the Moon July 1969, AD. We came in peace for all mankind.**
>
> **Plaque left on the Moon, 1969**

A body about the size of Mars – named Theia – supposedly collided with the Earth around 50 million years after the solar system formed, approximately 4.5 billion years ago. The impact was so forceful that it nearly shattered the fledgling Earth and the heat generated caused the upper layers of both bodies to melt. Theia's heavy iron core fell in to join the Earth's, and the lighter mantle and crust of the Earth was splashed off into space. It was this material that coalesced to produce the Moon.

1975	**1996**	**2009**
Giant impact hypothesis developed	Clementine spies water on Moon	LCROSS and Chandrayaan confirm water

Tides and orbital locking

The Moon turns the same face towards us each night. This is because it revolves about the Earth in the same time it takes to spin on its axis (about 29 days). This synchrony arises due to the effects of tides. The Moon's gravity distorts the fluid surface of the Earth, drawing out a bulge in the oceans towards the satellite itself, and equally on the planet's opposite side. These swelling bulges are responsible for the tides, which change as the Moon goes around the Earth. But they also act as locks on the Moon: if the planet and the Moon spin at different rates, the gravitational effect of the bulge will tug the Moon back into step.

The giant impact hypothesis explains why the Moon is so large relative to the Earth, yet has a tiny iron core. The lower average density of the Moon (3.3 gram/cubic centimetre) compared with the Earth (5.5 g/cc) results because the Moon lacks heavy iron. Moon rock also has exactly the same ratios of various oxygen isotopes (heavier radioactive versions of oxygen) as Earth, implying that it formed in the same neighbourhood. Martian rocks and meteorites formed in other parts of the solar system, in contrast, have very different compositions. Computer simulations of the mechanics of the impact confirm that the scenario is plausible.

Further evidence points to the fact that the Moon's surface was once molten – forming a magma ocean. Light minerals have floated to the surface of the Moon, as would be expected if they crystallized within a liquid phase. That the surface cooled slowly, perhaps taking as long as 100 million years to solidify, is indicated by the quantities of various radioactive isotopes, whose decay times can be used to measure mineral ages. There are some inconsistencies – the Moon has different ratios of volatile elements, as well as its deficit of iron, relative to the Earth. Similarly, there is no sign of Theia itself, in the form of unusual isotopes or foreign rock remnants. There's no smoking gun.

Differentiation As the Moon cooled, minerals crystallized out of the magma ocean and settled at depths according to their weight. The body differentiated to form a light crust, intermediate mantle and heavy core. The crust, just 50 km thick, is rich in lightweight minerals, including plagioclase (a feldspar found in granite). It is composed by mass of around 45 per cent oxygen and 20 per cent silicon, with the rest being made up of

> **And from my pillow, looking forth by light
> Of moon or favouring stars, I could behold
> The antechapel where the statue stood
> Of Newton with his prism and silent face,
> The marble index of a mind for ever Voyaging
> through strange seas of Thought, alone**

William Wordsworth

metals including iron, aluminium, magnesium and calcium. The core is small, confined to within 350 km or less. It is likely to be part melted and rich in iron and metals.

In between is the mantle, which undergoes moonquakes as it is twisted by tidal forces. Although thought to be solid now, over the lifetime of the Moon it has been molten and generated volcanism up until a billion years ago. The Moon's surface is scarred with numerous craters from impacts, which have scattered rocks and dust across its surface, in a layer known as regolith.

Water The Moon's surface is dry, but the occasional crash of comets or icy bodies on to its surface may have brought water with them. For lunar exploration, as well as for learning about the transference of material throughout the solar system, it is important to know if there is any water on the Moon or not. Whilst water would quickly evaporate in direct sunlight, there are some parts of the Moon that are in permanent shadow, especially on the sides of craters near the poles. Physicists suspect that water ice may survive in these dark locations.

Numerous orbiting satellites have scoured the surface, with mixed results. Polar water ice was reported by the Clementine and Lunar Prospector satellites in the late 1990s, although ground-based radio observations failed to confirm it. Recent missions, including NASA's LCROSS (Lunar Crater Observation and Sensing Satellite), in which a projectile was fired into the surface and onboard instruments analysed the light from the resulting plume, and India's Chandrayaan mission, claim to have detected water in the shadows of craters. So future astronauts might find enough to drink on the Moon's parched surface.

the condensed idea
One small step

49 Astrobiology

Life flourishes on Earth. Historically we have long believed that life exists beyond our planet – from Martian canals to reports of flying creatures on the Moon. But the more we have probed our solar system, the more barren our neighbourhood seems. Although life is robust, it seems to need particular conditions to get going. Astrobiology seeks to answer the question of how life arises in the cosmos, and where.

Life began on Earth very soon after the planet formed, 4.5 billion years ago. Fossil stromatolites – dome-shaped organic mats – show that cyanobacteria were in existence 3.5 billion years ago. Photosynthesis – the chemical process that uses sunlight to convert chemicals into energy – was also well under way. The oldest known rocks, identified in Greenland, are 3.85 billion years old. So there is a small window in which life kick-started.

Theories of the origin of life are as old and as diverse as species. Micro-organisms such as bacteria and protozoa were first spotted in the 17th century, when the microscope was invented. The apparent simplicity of bacteria led scientists to suppose that the blobs had grown spontaneously from inanimate matter. But then they were seen to replicate, suggesting that life was self-generating. In 1861, Louis Pasteur tried and failed to create bacteria from sterile nutrient-rich broth. Building the first organism was problematic.

Charles Darwin addressed the issue of the origin of life in a letter to botanist Joseph Hooker in 1871: it may have begun in a 'warm little pond, with all sorts of ammonia and phosphoric salts, lights, heat, electricity, etc. present, so that a protein compound was chemically formed ready to undergo still more complex changes'.

timeline

1861
Louis Pasteur fails to create
life from a nutrient broth

1871
Charles Darwin ponders
his 'warm little pond'

> **❝It is no valid objection that science as yet throws no light on the far higher problem of the essence or origin of life. Who can explain gravity? No one now objects to following out the results consequent on this unknown element of attraction ...❞**
>
> **Charles Darwin**

Primordial soup Darwin's explanation is close to what scientists believe today, with one important addition. Lacking plants and biological sources of oxygen, the early Earth's atmosphere was oxygenless, unlike today. It contained methane, ammonia, water and other gases that favoured certain types of chemical reactions over others. In 1924 Alexander Oparin suggested that under these conditions a 'primeval soup of molecules might have been created'. Those same processes would be prevented from taking place now in our oxygen-rich atmosphere.

Conditions on the early Earth were hell-like – as reflected in the geological name given to the epoch, the Hadean period. Coming into existence as soon as 200 million years after the Earth's formation, the oceans were initially boiling hot and acidic. The late heavy bombardment was ongoing, so asteroids will have frequently crashed to the planet's surface. Turbulent weather, including electric storms and deluges of rain, made the place inhospitable. Yet these conditions might have been conducive to life. The myriad organisms that live around deep sea vents on the marine floor show that boiling water and darkness are no barrier, providing there are enough nutrients. Even so, the first organisms had to develop somehow from complex molecules.

The hostile conditions of the early Earth might have been suited to creating organic molecules. Laboratory experiments in 1953 by Stanley L. Miller and Harold C. Urey showed that small molecules basic to life, such as amino acids, can be created from a mixture of gases – methane, ammonia and hydrogen – by passing electricity through it. Scientists have got little

1950s	**1953**	**2005**	**2020**
Fred Hoyle promotes 'panspermia'	Miller–Urey experiment	Huygens probe lands on Titan	Launch of Europa mission

Huygens probe

The Huygens space probe landed on the surface of Titan on 14 January 2005, after a seven-year journey. Contained inside a protective outer shell a few metres across, it carried a suite of experiments that measured the winds, atmospheric pressure, temperature and surface composition as it descended through the atmosphere to land on an icy plain. Titan is a weird world whose atmosphere and surface is damp with liquid methane. Huygens was the first space probe to land on a body in the outer solar system.

further in the decades since then, however. The architectural step of building the first cells from molecules is challenging: pod-like structures that may be formed by lipids are one suggestion for a precursor. The processes of cell division and setting up of a chemical engine – metabolism – are still far from understood. No one has yet made a convincing proto-cell from scratch.

Panspermia An alternative possibility is that the complex molecules, and perhaps simple biological organisms, originated in space. Around the same time that the Miller–Urey experiment was performed, astronomer Fred Hoyle was pushing the idea of 'panspermia' – proposing that life on Earth was seeded by meteorite and comet impacts. Although this might seem far-fetched, space is full of molecules, some of them complex. The amino acid glycine was detected in 2009 in material ejected from Comet Wild-2 that was sampled and brought back to Earth by NASA's Starburst probe.

To learn more about conditions that might support early forms of life, and how molecules might have spread, astrobiologists are keen to explore key places in our solar system. Mars is the primary target. Although its surface is dry today, it is thought to have been wet in the past. Water ice lingers at its icy poles and Mars Rover images have revealed evidence that liquid water has flowed at its surface, perhaps in small streams or due to a fluctuating subterranean water table. Methane has been detected in the red planet's atmosphere, suggesting a geological or perhaps biological origin.

Astrobiology tourism Saturn's largest moon, Titan, is another location that may be conducive to life, and bears similarities to early Earth. Although located in the frozen outer solar system, it is swathed in a thick

> ❝The great age of the Earth will appear greater to man when he understands the origin of living organisms and the reasons for the gradual development and improvement of their organization.❞
>
> **Jean-Baptiste Lamarck**

nitrogen atmosphere that contains many organic molecules, including methane. The Moon was visited in 2005 by a probe dropped from NASA's Cassini spacecraft, which is investigating Saturn. The capsule, named Huygens after the 17th century Dutch physicist who discovered the moon, descended through the clouds of Titan's atmosphere to land on its surface of frozen methane. Titan has continents, sand dunes, lakes, and perhaps rivers, made of solid and liquid methane and ethane, rather than water. Some think it could harbour primitive life forms such as methane-eating bacteria.

Another of Saturn's moons, Enceladus, is a popular astrobiology destination. As the Cassini probe flew past this ice-covered moon, it spotted a vast plume of water vapour coming from cracks near its south pole. A warm spot below is releasing steam through vents, broken open as the moon is twisted by tidal forces generated by its proximity to Saturn. It is possible that life could survive beneath the surface where there is liquid water.

The most likely destination for the next focused astrobiology mission is Jupiter's moon Europa, which harbours a liquid water ocean beneath its frozen surface. Like Enceladus, its surface is smooth, indicating that it has recently been melted. Fine cracks stripe it, suggesting that it gains heat too through tidal flexing. Life might be sheltered within this ocean, paralleling conditions in the Earth's deep seas and buried ice lakes in Antarctica. Astrobiologists plan to send a space mission to Europa in 2020 to drill through its ice and search for signs of life.

the condensed idea
Follow the water

50 Fermi paradox

The detection of life elsewhere in the universe would be the greatest discovery of all time. Physics professor Enrico Fermi wondered why, given the age and vastness of the universe, and the presence of billions of stars and planets that have existed for billions of years, we have not yet been contacted by any other alien civilizations. This was his paradox.

Chatting with his colleagues over lunch in 1950, Fermi supposedly asked, 'Where are they?' Our own galaxy contains billions of stars and there are billions of galaxies in the universe, so that is trillions of stars. If just a fraction of those anchored planets, that's a lot of planets. If a fraction of those planets sheltered life, then there should be millions of civilizations out there. So why haven't we seen them? Why haven't they got in touch with us?

Drake equation In 1961, Frank Drake wrote down an equation for the probability of a contactable alien civilization living on another planet in the Milky Way. This is known as the Drake equation. It tells us that there is a chance that we may coexist with another civilization but the probability is still quite uncertain. Carl Sagan once suggested that as many as a million alien civilizations could populate the Milky Way, but he later revised this down, and others since have estimated that the value is just one, namely humans.

More than half a century after Fermi asked his question, we have still heard nothing. Despite our communication systems, no one has called. The more we explore our local neighbourhood, the lonelier it seems. No concrete signs of any life, not even the simplest bacteria, have been found on the

timeline

AD1950	1961
Fermi questions the absence of alien contact	Drake devises his equation

Moon, Mars, asteroids or the outer solar system planets and moons. There are no signs of interference in the light from stars that could indicate giant orbiting machines harvesting energy from them. And it is not because no one has been looking. Given the stakes, there is great attention paid to searching for extraterrestrial intelligence.

Search for life How would you go about searching for signs of life? The first way is to start looking for microbes within our solar system. Scientists have scrutinized rocks from the Moon, but they are inanimate basalt. It has been suggested that meteorites from Mars might contain the remnants of bacteria, but it is still not proven that the ovoid bubbles in those rocks hosted alien life and were not contaminated after having fallen to Earth or produced by natural geological processes. Cameras on spacecraft and landers have scoured the surfaces of Mars, asteroids and now even a moon in the outer solar system – Titan, orbiting Saturn. But the Martian surface is dry and Titan's surface is drenched in liquid methane, though so far devoid of life. Jupiter's moon Europa may host seas of liquid water beneath its frozen surface. So liquid water may not be so rare a commodity in the outer solar system, raising expectations that one day life may be found.

But microbes are not going to call home. What about more sophisticated animals or plants? Now that individual planets are being detected around distant stars, astronomers are planning on dissecting the light from them to hunt for chemistry that could support or indicate life. Spectral hints of ozone or chlorophyll might be picked up, but these will need precise

Outside intelligences, exploring the solar system with true impartiality, would be quite likely to enter the Sun in their records thus: star X, spectral class G0, 4 planets plus debris.
Isaac Asimov, 1963

1996
Antarctic meteorites hint at
primitive life existing on Mars

observations, like those possible with the next generation of space missions such as NASA's Terrestrial Planet Finder. These missions might find us a sister Earth one day, but if they did, would it be populated with humans, fish or dinosaurs, or just contain empty, lifeless continents and seas?

Contact Life on other planets, even Earth-like ones, might have evolved differently to that on Earth. It is not certain that aliens there would be able to communicate with us. Since radio and television began broadcasting, their signals have been spreading away from Earth, travelling outwards at the speed of light. So any TV fan on Alpha Centauri (four light years away) would be watching the Earth channels from four years ago, perhaps enjoying repeats of the film *Contact*. Black and white movies would be reaching the star Arcturus, and Charlie Chaplin could be starring at Aldebaran.

Earth is giving off plenty of signals, if you have an antenna to pick them up. Wouldn't other advanced civilizations do the same? Radio astronomers are scouring nearby stars for signs of unnatural signals. The radio spectrum is vast, so they are focusing on frequencies near key natural energy transitions, such as those of hydrogen, which should be the same anywhere in the universe. They are looking for transmissions that are regular or structured but are not made by any known astronomical objects.

Drake equation

$$N = N^* \times f_p \times n_e \times f_l \times f_i \times f_c \times f_L$$

where:

N is the number of civilizations in the Milky Way Galaxy whose electromagnetic emissions are detectable

N^* is the number of stars in the galaxy

f_p is the fraction of those stars with planetary systems

n_e is the number of planets, per solar system, with an environment suitable for life

f_l is the fraction of suitable planets on which life actually appears

f_i is the fraction of life-bearing planets on which intelligent life emerges

f_c is the fraction of civilizations that develop a technology that releases detectable signs of their existence into space

f_L is the fraction of a planetary lifetime such civilizations release detectable signals into space (for Earth this fraction is so far very small).

> **Our Sun is one of 100 billion stars in our galaxy. Our galaxy is one of billions of galaxies populating the universe. It would be the height of presumption to think that we are the only living things in that enormous immensity.**
>
> Werner von Braun

In 1967, English graduate student Jocelyn Bell got a fright in Cambridge when she discovered regular pulses of radio waves coming from a star. Some thought this was indeed an alien Morse code, but in fact it was a new type of spinning neutron star now called a pulsar. Because this process of scanning thousands of stars takes a long time, a special programme has been started in the USA called SETI (Search for Extra-Terrestrial Intelligence). Despite analysing years of data, the programme has not yet picked up any odd signals. Other radio telescopes search occasionally, but these too have seen nothing that does not have a more mundane origin.

Out to lunch So, given that we can think of many ways to communicate and detect signs of life, why might any civilizations not be returning our calls or sending their own? Why is Fermi's paradox still true? There are many ideas. Perhaps life only exists for a very short time in an advanced state where communication is possible. Why might this be? Perhaps intelligent life always wipes itself out quickly. Perhaps it is self-destructive and does not survive long, so the chances of being able to communicate and having someone nearby to communicate to are very low indeed. Or there are more paranoid scenarios. Perhaps aliens simply do not want to contact us and we are deliberately isolated. Or perhaps they are just too busy and haven't got around to it yet.

the condensed idea
Is there anybody out there?

Glossary

Absolute zero a temperature of −273 degrees Celsius; the coldest temperature it is possible to attain.

Absorption line Gap in spectrum at a particular frequency of light.

Acceleration change in something's velocity in a given time.

Active galaxy a galaxy exhibiting high-energy processes at its centre driven by a supermassive black hole.

Age of universe about 14 billion years old, determined from rate of expansion.

Atom smallest building block of matter that can exist independently.

Baryon particles made up of electrons, protons and neutrons.

Black body radiation light glow emitted by a black object at one temperature.

Black hole gravitationally extreme region from which light cannot escape.

Cepheid variable star whose period scales with its luminosity.

Constellation recognized pattern of stars on the sky.

Cosmic microwave background faint microwave glow coming from all over the sky from the Big Bang.

Dark energy form of energy in empty space that causes spacetime to expand.

Dark matter invisible material detectable only by its gravity.

Diffraction spreading out of waves when they pass a sharp edge or slit.

Dust in cosmos, soot and particulates that absorb and redden light.

Electromagnetic wave transmits energy through electric and magnetic fields.

Emission line brightening of a specific light frequency in a spectrum.

Energy quantity that dictates potential of change by being exchanged.

Exoplanet planet orbiting a star other than the Sun.

Field magnetic, electric, gravity – a means of transmitting a force at a distance.

Fission the splitting apart of heavy nuclei into lighter ones.

Force lift, push or pull that changes something's motion.

Frequency the rate at which wave crests pass some point.

Fusion the combination of light nuclei to make heavier ones.

Galaxy a defined grouping of millions of stars, such as our Milky Way.

Gas a cloud of unbound atoms or molecules.

Gravitational lensing the bending of light rays as they travel past a massive object.

Gravity a fundamental force by which objects attract one another.

Hubble constant rate of expansion of the universe.

Inertia – see mass.

Inflation very rapid swelling of universe in first fraction of a second.

Interference the combining of waves of different phases that may reinforce or cancel out each other.

Ion an atom with electric charge due to the loss or addition of an electron.

Isotope element forms with differing nuclear masses due to extra neutrons.

Isotropy uniform distribution of something, evenly spread.

Light elements first few elements formed in big bang: hydrogen, helium, lithium.

Mass attributed to number of atoms or equivalent energy in something.

Molecule combination of atoms held together by chemical bonds.

Momentum the product of mass and velocity that expresses how hard it is to stop something once moving.

Multiverse system of many parallel but separate universes.

Nebula a fuzzy cloud of gas or stars; early name for galaxy.

Neutron star a collapsed husk of a burned out star, supported by quantum pressure.

Nucleus hard central core of an atom made of protons and neutrons.

Nucleosynthesis the formation of elements by nuclear fusion.

Orbit the ring-like path of a body, often elliptical.

Phase the relative shift in wavelength between the peaks of two waves.

Photon light manifesting as a particle or packet of energy.

Planet a self-gravitating, orbiting body too small to undergo fusion.

Pressure force per unit area.

Pulsar a spinning, magnetized neutron star that sends out radio pulses.

Quantum mechanics the laws of the sub-atomic world, many of which are counterintuitive but follow mathematical rules.

Quantum pressure a fundamental limit driven by the rules of quantum mechanics that prevents some types of particles from existing in identical states in close proximity.

Quark a fundamental particle, three of which combine to make up protons and neutrons.

Redshift frequency drop of a receding object due to expansion of universe.

Reflection reversal of a wave as it hits an impenetrable surface.

Refraction bending of waves, such that they slow down in denser media.

Spacetime geometric space combined with time in one function in relativity.

Spectrum sequence of electromagnetic waves, from radio to gamma rays.

Standard model accepted theory of families of fundamental particles.

Star a ball of gas undergoing fusion in its core.

Supernova explosion of a dying star when fusion stops.

Supermassive black hole a black hole with mass equivalent to millions of stars.

Temperature in Kelvin, measured relative to absolute zero (−273 °C).

Universe all of space and time, by definition including everything.

Vacuum empty space, containing no atoms; outer space isn't completely empty.

Velocity speed in a particular direction.

Wavelength distance between the crests of a wave.

Wave-particle duality behaviour, particularly of light, that that is sometimes wave-like and at other times like a particle.

Index

Quercus Publishing Plc
21 Bloomsbury Square
London WC1A 2NS

First published in 2010

A catalogue record of this book is available from
the British Library

UK and associated territories:
ISBN 978 1 85738 123 1
US and associated territories:
ISBN 978 1 84866 066 3

Printed and bound in China

10 9 8 7 6 5 4 3 2 1

Prepared by Starfish Design, Editorial and
Project Management Ltd.